"VICTORIA"

A 19th Century Steam Engine

Constructional Details

by

Andrew Smith

Model and Allied Publications
Argus Books Ltd.
14 St. James Rd., Watford
Herts, England

ISBN 0 85242 773 5

First published 1977
Revised edition 1982

Printed in England by Staples Printers Rochester Limited at The Stanhope Press.

Preface

Although the Stuart Turner series of model steam engines covers a very wide range of types, there is always the interesting possibility of developing a new configuration by making use of existing castings. The accuracy of dimension and easy working for which these castings are noted makes such a choice even more attractive in a hobby where "one off's" tend to be difficult to obtain. Such "variations" serve a number of useful purposes; they form an interesting armchair and drawing board exercise and by increasing the use of such castings make their manufacture more economic at a time when any new production is expensive.

The proportions of the cylinder, flywheel and other castings used in this series of engines, belongs to a period when there seemed to be a limitless range of variations in steam engine styles. Apart from the Beam and the Horizontal, there was the Table Engine, the Grasshopper, the Side Lever, the Bell Crank and the Vertical. In fact, the type of engine that today we refer to as the Vertical would, in those days, have been considered an Upside-down Vertical!

I hope that you will enjoy building and steaming these engines, they will not power your workshop, but I am certain will reward you with a great deal of pleasure and quiet contentment.

ANDREW SMITH.

Saltford,
Avon,
England.

Contents

Page

1. How it all began and an introduction to "Victoria" 6

2. Baseplate, pedestal, bearings, crankshaft and flywheel 11

3. The cylinder and all its parts 20

4. Connecting rod, crosshead and guides, valve gear 30

5. A woodworking session, erecting the engine 41

6. Developing the "Victoria" Engine 47

7. A Double-Size "Victoria" Mill Engine 55

8. Camden Industrial Museum 59

Building the "VICTORIA"
A 19th Century Steam Engine
Constructional Details

by

Andrew Smith

The completed "Victoria"

A 19th Century Steam Engine

1. How it all began and an introduction to "Victoria"

At the corner of two streets in a fair city in what was the County of Somerset, stood an establishment, the like of which will never again be seen—Bowlers of Bath. The outside proclaimed that it was an engineers, blacksmiths, brass founders and maker of mineral waters; the inside had not altered since the gaslit days of Queen Victoria. To anyone interested in the history of industry, the building was brimful of power units and machinery. In one small room, a triangular bed lathe with all its tools—just as the craftsman had left it; in various other parts of the premises there remained still the engines—steam, gas, oil—that had once powered the machinery. This establishment is no more, it has been demolished, to make way for—a car park!

One day, while talking steam to the quiet spoken, white haired owner, he mentioned that he thought the original steam engine, installed probably nearly a hundred years before, was still in the building and that if I was prepared to remove it, I was welcome to it. The result was that a friend and I presented ourselves a few days later in old clothes and with spanners and crowbars, to find that the engine was on a first floor, but that the surrounding walls had been removed, leaving the floor carrying the engine as a kind of shelf, the only access to which was via a ladder.

The engine was a horizontal of mid-19th century type, lightly built for the low pressures of the time, and having a bore of 5 inches and a stroke of about 10 inches; spindly connecting rod and large flywheel. After a great deal of work during which we sampled the establishment's prowess as "makers of mineral waters," we got the engine dismantled and out of the building.

After cleaning, painting, re-machining the port face and slide valve, and rebuilding the crankshaft and its bearings, the engine was shown to Mr. George Watkins, author of "The Stationary Steam Engine" and a member of SIMEC, who suggested that it might have been built by a long defunct local firm about 1860. At this time, I decided that the engine took up too much room in my workshop and as I was never likely to steam it, disposed of it to a gentleman who was setting up a museum of power units—he had everything from pony mills to fractional horse-power electric motors.

For a long time I have given thought to a model of this engine. Recently when looking at photographs of the Beam Engine in the Stuart Turner catalogue, I suddenly realised that the proportions of the cylinder, connecting rod and flywheel were exactly similar to this workshop engine, but of course, to a different scale. Out came the drawing board and the result you see in Fig. 0. While it is not exactly a model of the particular engine with which I was concerned, it is nevertheless closely representative of this type of workshop power unit. The ratio of bore to stroke, the very

A photograph taken in the 1920*s of the Bowler Works. The machinery was originally driven by the locally built horizontal steam engine on which the "Victoria" is based. The premises were demolished during* 1970 *to make way for a car park.*

small steam chest, thin connecting rod, all mounted on a wooden plinth, supplied with steam at no more than 50 lbs. pressure and running at a mere hundred or so revolutions per minute, everything was indicative of the period.

For the newcomer to model steam engine building, with some small experience of simple lathe and bench work, this type of model offers advantages. It is what might be described as a "roomy" model, in that the various sub-assemblies, cylinder, crosshead guides, crankshaft, are well spread out and offer room for adjustment on assembly to allow for any slight errors in workmanship. It is a good looking model, with all parts visible, which is the charm of the old fashioned steam engine, and because, in practice it would be expected to run comparatively slowly on low pressure steam, only a small boiler will be needed to steam it for demonstration purposes. In fact, any of the Stuart Turner Babcock Boiler range would be suitable.

In spite of the full size version having a cylinder bore of 5 inches and a stroke of 10 inches, it would only have been rated at $2\frac{1}{2}$ horsepower; an extremely conservative figure. The early method of calculating horsepower on such steam engines was to divide the square of the cylinder bore by 10, i.e. $hp = bore^2 \div 10$. With a bore of 5 inches this gives a

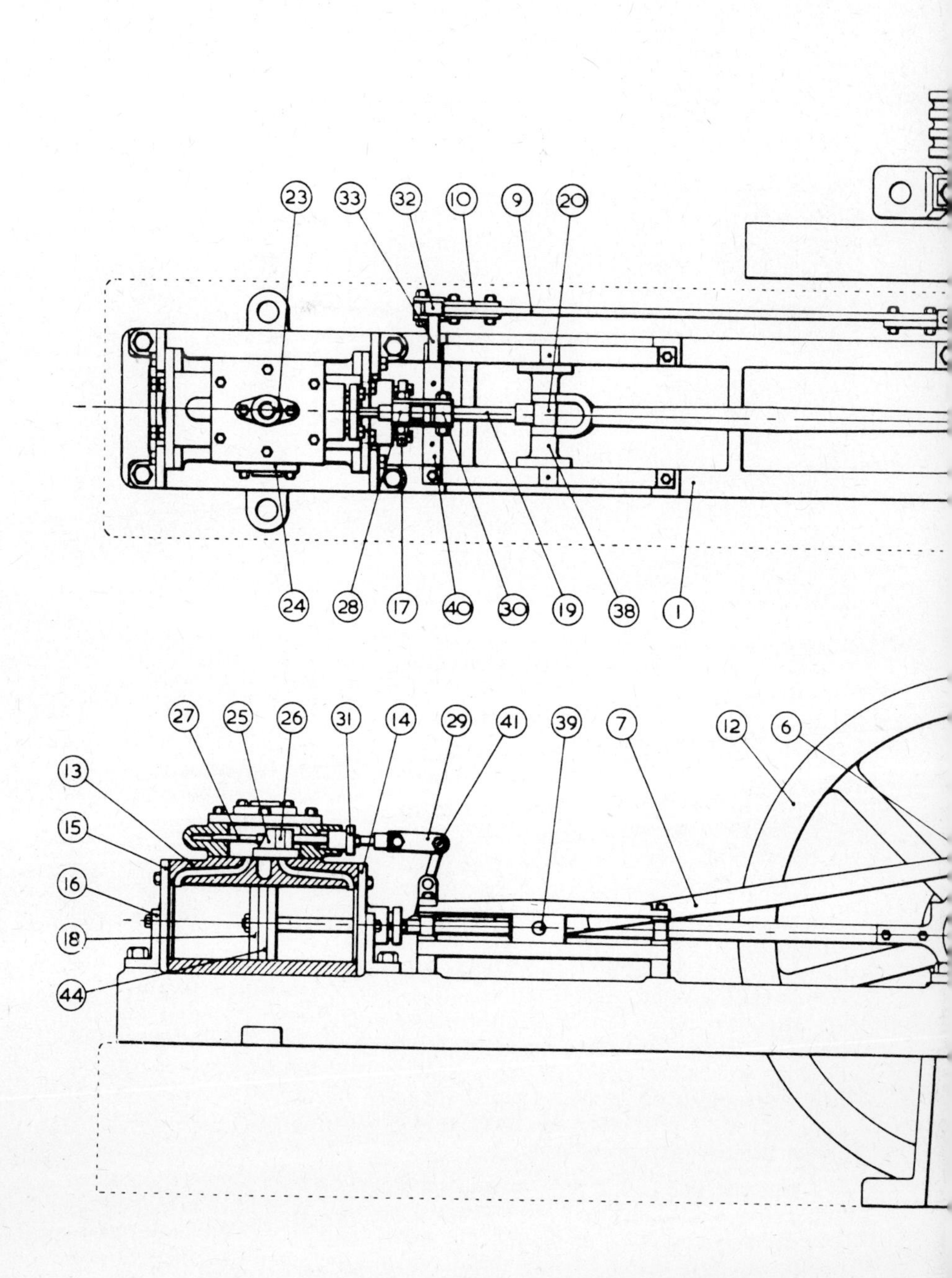
23
33
32
10
9
20
24
28
17
40
30
19
38
1
27
25
26
31
14
29
41
39
7
12
6
13
15
16
18
44

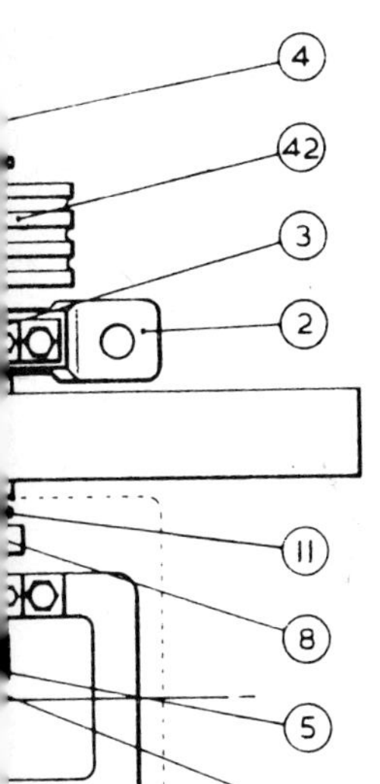

THE VICTORIA ENGINE

FIG. 0. *These general arrangement views are numbered to correspond with the part numbers used in the Stuart drawings and List of Parts, they are also used as Fig. Nos. throughout the book.*

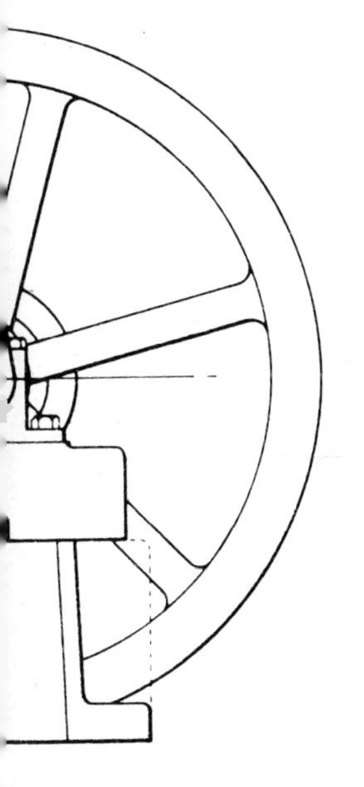

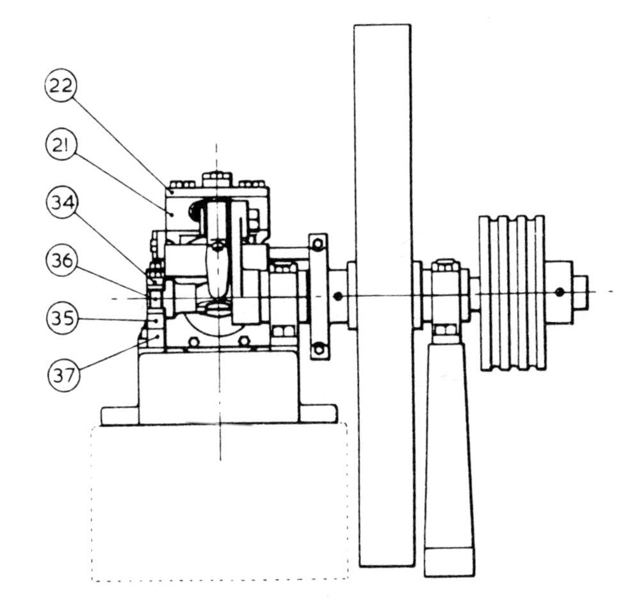

The "Victoria" follows closely the design of the low-pressure, horizontal engines used in the mid-19th century for workshop and factory power. The model is 1 *inch bore by* 2 *inch stroke. Length:* 15½ *inches* (394 *mm*). *Width:* 6½ *inches* (165 *mm*). *Height:* 7⅛ *inches* (181 *mm*). *Finished weight:* 11½ *lbs.* (5.2 *kg*).

horsepower of 2½. Notice that there is no reference whatever to steam pressure, revolutions per minute or stroke. Using the same method of calculation our proposed model could be described as 'one tenth horsepower.'

If you are a high speed merchant who thinks of engine speeds in terms of thousands of revs per minute, this engine will not appeal to you. But if an open fire, a glass of whisky and an engine whispering over so slowly that every movement can be clearly seen interests you, then build "VICTORIA," and, given the above conditions, I'll join you for her initial steam test . . .!

Before we start work, I should point out that most of the drawings in this book are those made when initially designing and building the prototype of this model engine and may vary in unimportant detail with the drawings supplied to customers by Stuart Turner Ltd. Where the builder may have any doubts as to the correct dimension to use he is advised, especially if a comparative beginner, to stick strictly to those on the "works" drawings, i.e. those that came with the set of castings.

Finally, where the drawing shows a tapped hole with a second dimension, e.g. 2BA x 7/16, the 7/16 refers to the required depth of the hole.

2. Baseplate, pedestal, bearings, crankshaft and flywheel

Although the baseplate for this engine is large, some 13½ inches long and over 3 inches wide, the accuracy of the casting is such that it should cause no problems at all to the model engineer with limited workshop equipment. In fact all the facing of the surfaces may be easily accomplished using file and emerycloth. A strip of the latter of a fine grade, stretched along a flat surface, or even tacked on an 18 inch length of stout timber which has been carefully planed, will, in a short time, produce a satisfactory surface, after the initial roughness has been removed with a smooth flat file.

The 1 inch height of the baseplate is not fussy, and even the need for parallelism between bottom and top is not critical as it can be adjusted by slight modifications to the wooden plinth and/or the use of shims. Do not imagine for a moment that I am condoning slip-shod work, but like the craftsmen of 150 years ago, we are doing what may seem to be the impossible, with the minimum of simple tools.

Mark out and centre punch the holes in the bolting down lugs. Drill a pilot hole and open out with a 17/64 inch drill to take ¼ inch holding down bolts. Then spot face these lugs with a ¼ inch pin drill or clean them up with a file if such a large spot facing cutter is not available. It is an interesting bit of toolmaking to make these from a short length of ½ inch silver steel. You will notice that the bolting lugs are not symmetrical about the casting. They actually come where they are most needed, i.e. in the centre of the cylinder block location and at the crankshaft.

Next scribe the crankshaft centre-line across the surface where the bearing will be mounted, and from this centre-line mark out, drill and tap 2BA the two holes for the bearing. Using the centre-line again as reference, mark out and lightly dot-punch the four 2BA holes which hold the cylinder and likewise the four 7BA holes for the crosshead guide bars. Thus with a total of six 2BA and four 7BA tapped holes the baseplate is complete.

You may feel doubtful of your ability to work accurately enough to be sure that the tapped holes in the baseplate will be in the right positions when it comes to the final assembly. In which case you may be tempted to leave this part of the job and complete it later by spotting through from the various other components. There is nothing wrong with this method of course, but it might help your skill and confidence to have a few practice "marking out" sessions on some odd bits of plate until you get the feel of it, then you can tackle jobs of this sort with confidence.

Obviously the right marking out equipment is needed, and most of this you will probably buy. Get into the habit of working from a datum surface, the lathe bed or faceplate, a piece of plate glass, or a marble washbasin top. Use short end of bright steel bar to help, it is accurate to within a few thousandths of an inch—far more accurate than you can achieve using a steel rule. Make sure the scriber has a long tapered point and is really sharp. but don't make one from a long piece of ⅛ inch silver steel. This is much too uncomfortable to handle. Use about 2 inches of silver steel for the end and mount it in a handle of

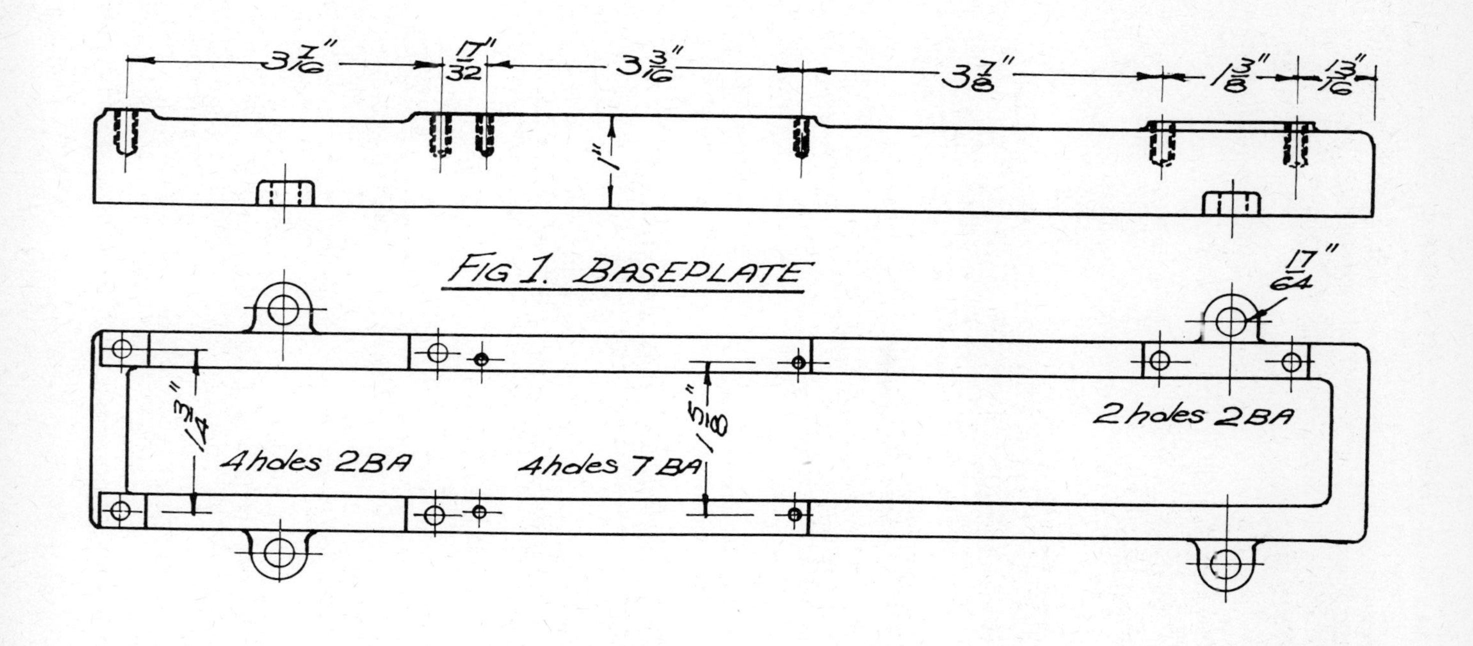

FIG 1. BASEPLATE

Main bearing set up in the four-jaw chuck, about to be reamed

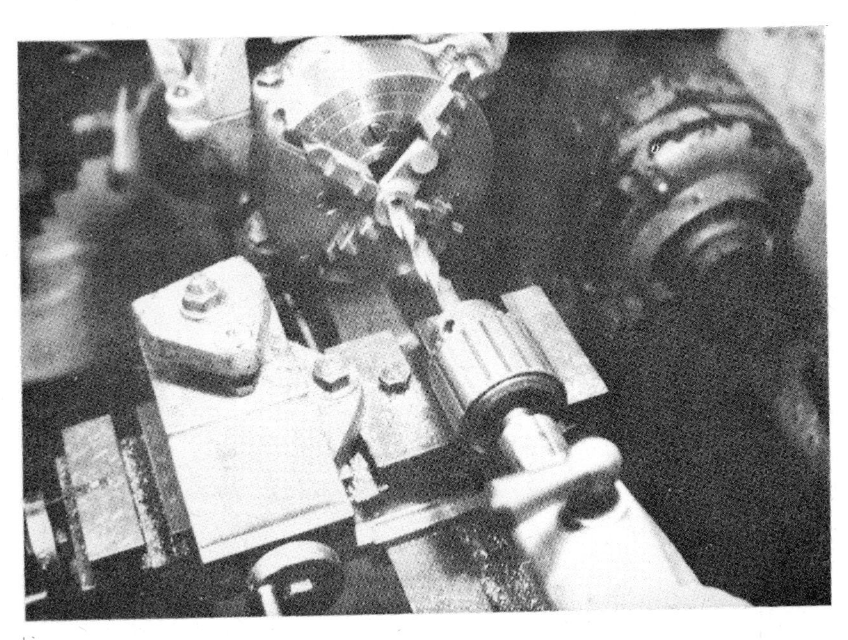

Drilling the crankweb prior to reaming

5/16 or $\frac{3}{8}$ inch diameter so that you can hold it really firmly. Two centre punches are needed, one having a point of less than 60° included angle, which should be so sharp that it will drop into the junction of two scribed lines. The other with a 90° angle is used for enlarging the dots prior to drilling.

With the baseplate complete, tackle the pedestal (part No. 2) next. This again, although a fairly large component, requires only the minimum of work to be carried out on it. The top and bottom surfaces may be filed flat and to 3 inches apart, or it may be set up on the lathe faceplate as shown in Fig. 2a. As you will see from this sketch, location is from the central web of the pedestal, via a packing piece. When the job is reversed to face the other end be sure that you do not inadvertently turn it over. Light cuts are the order of the day, with such a large overhang. Mark out, drill and tap where necessary then file a clearance in the base as shown, and another part of the engine structure is complete.

The main bearings are made of top and bottom castings in gunmetal. Clean these with a file, then mount in the four-jaw chuck and face the mating surfaces on the four parts. Heat them with a blowpipe or on a gas ring and sweat over the machined surfaces with soft solder. Solder paint containing its own flux is ideal for this sort of work. While hot, wipe off the excess solder then place the top part of the bearing on the lower part and apply a slight pressure with a bit of stick to squeeze out the excess solder. Keep on the pressure until the solder has set—a little patience is needed here—then clean off any excess solder with an old file.

Mount each bearing in the four-jaw chuck, base outwards, and face off until the distance from joint to underside is $\frac{5}{8}$ inch. Remount in an upright position in the chuck with the bearing centre running true. Face, then centre drill, open out and ream to 7/16 inch or finish to size with a small boring tool if a reamer is not available. With a knife tool face down the side of the bearing block leaving a protruding rim 11/16 inch diameter and $\frac{1}{8}$ inch long. Slacken two chuck jaws only—not the one supporting the bottom of the bearing—and reverse the bearing. Reset, so that the 7/16 inch bore is running true, face to an overall thickness of $\frac{5}{8}$ inch, then reduce the side of the bearing to $\frac{3}{8}$ inch thick leaving a rim 11/16 inch diameter by $\frac{1}{8}$ inch long. Remove from the lathe and clean up the remainder of the bearing to size by careful filing. Mark out for the mounting holes at $1\frac{3}{8}$ inch centres, pilot drill and finish with a 2BA clearance drill. On the top cap of the bearing mark out for the 5BA bolt holes and drill tapping size to a depth of $\frac{5}{8}$ inch, spot-face the top surfaces. Now drill and countersink a 1/16 inch oil hole in the centre of the top bearing cap. Finally identify the pairs of bearings and unsweat them; clean off the solder, tap the lower halves of the clamping bolt holes 5BA and open out the bolt holes in the cap with a 5BA clearance drill.

If you have not already done them, spot through from the bearings for the 2BA holes on the baseplate and the pedestal. Drill them tapping size and carefully tap 2BA. Start the tap in the drilling machines or use a tap guide to ensure squareness of the thread in the holes.

With the bearings complete, it would be nice to have something to rotate in them, so let's make the crankshaft next. It is in three parts, the shaft, the crankweb and the crankpin. For reasons which will become

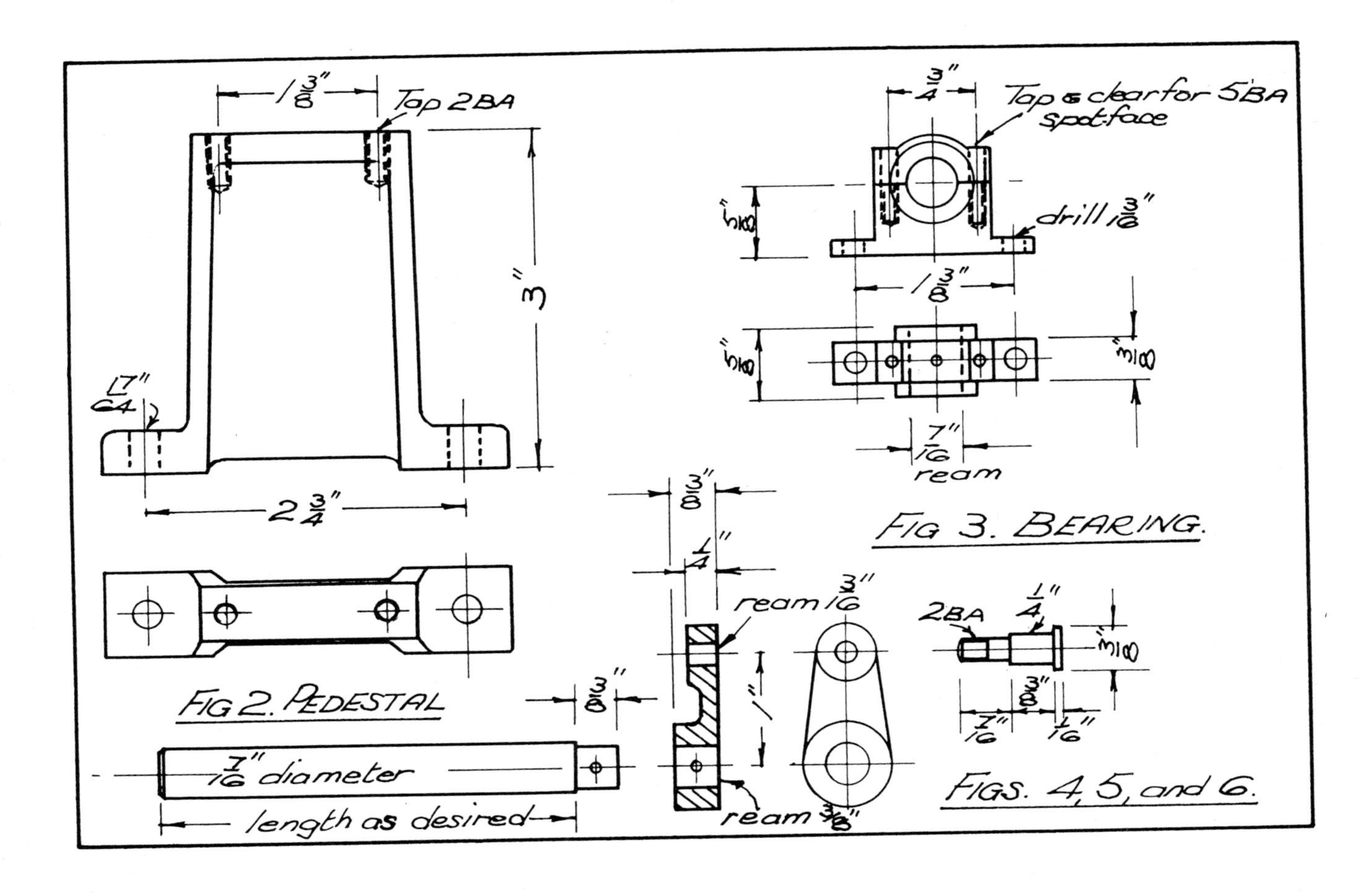

FIG 2. PEDESTAL

FIG 3. BEARING.

FIGS. 4, 5, and 6.

apparent later we make the web first. This is a neat little iron casting, Fig. 5, which initially requires cleaning with files to as near the finished shape as possible. Set in the four-jaw chuck with the flat side outwards and face off, making sure that you leave sufficient to allow for finishing to the required thickness. Reverse in the chuck, setting the $\frac{3}{4}$ inch diameter boss to run true. Face to $\frac{3}{8}$ inch thick, centre drill, drill and ream to $\frac{3}{8}$ inch. While the job is still mounted in the chuck, advance the tool and face down the small boss to $\frac{1}{4}$ inch thick. Remove from the lathe, mark out, centre punch, drill and ream the 3/16 inch hole in the smaller boss at 1 inch centres. If you do not have a drilling machine which is sufficiently accurate, I suggest you do this job with the crankweb clamped to the lathe faceplate and running true. Prior to reaming operations use a drill as near to the reaming size—but obviously below it—as possible. For example, for the 3/16 inch reamed hole, the drill would ideally have been No. 13 or 4·70 mm—which are actually both the same size—leaving only $2\frac{1}{2}$ thou to be removed by the reamer. On occasions when you do not have the correct size reamer follow the same procedure, but finish with a drill of the required size.

The next couple of jobs, the shaft (Fig. 4) and the crankpin (Fig. 6), are plain turning, so mount the self-centring chuck and a keen knife tool in the lathe. Face the 7/16 inch diameter bar to length and reduce one end to a light press fit in the $\frac{3}{8}$ inch hole in the crank web. Do this carefully because a too easy fit will require starting again and shortening the shaft, while a too tight fit may necessitate a letter to Henley-on-Thames for a spare crankweb! So, light cuts and check frequently. Drill and fit a 3/32 inch steel crosspin, use a No. 42 drill.

The crankpin is turned from $\frac{3}{8}$ inch steel bar. Mount in the chuck with sufficient protruding to complete the turning in one setting. Face the end, then turn down a $\frac{7}{8}$ inch length to just over $\frac{1}{4}$ inch diameter. Reduce $\frac{1}{2}$ inch to 3/16 inch diameter aiming for a very close fit in the crankweb. Now reduce 5/16 inch of this length to 0·185 inch diameter and thread 2 BA, guiding the diestock with the tailstock barrel. When the thread is complete, shorten the length of this portion to the 7/16 inch shown on Fig. 6, leaving a nicely radiused finish to the end of the thread. Reduce the remainder to an accurate $\frac{1}{4}$ inch diameter with as fine a surface finish as possible. Finally, part off, leaving a head 1/16 inch thick. Assemble to crankweb locking with a 2BA net and put to one side until we machine the flywheel.

The flywheel, Fig. 12, has the spindly spokes typical of the period which this engine represents. The first operation is to file the casting overall to remove any excresences and moulding joint lines. Because you may find it difficult to hold the casting in the chuck of a small lathe for turning, I suggest that we use the faceplate for this operation. The diagram, Fig. 12a, shows the method of mounting the flywheel. Two pieces of packing, as shown, are needed between the spokes and the faceplate. Don't be tempted to do without this or you may hear an ominous crack as you tighten the clamping nuts! Set the wheel to run as truly as possible, then with the lathe set on its lowest speed, face the side and periphery of the rim and the boss. On mine I found that only two "10 thou" cuts were needed on the rim to find good metal, so I left it at that. A little

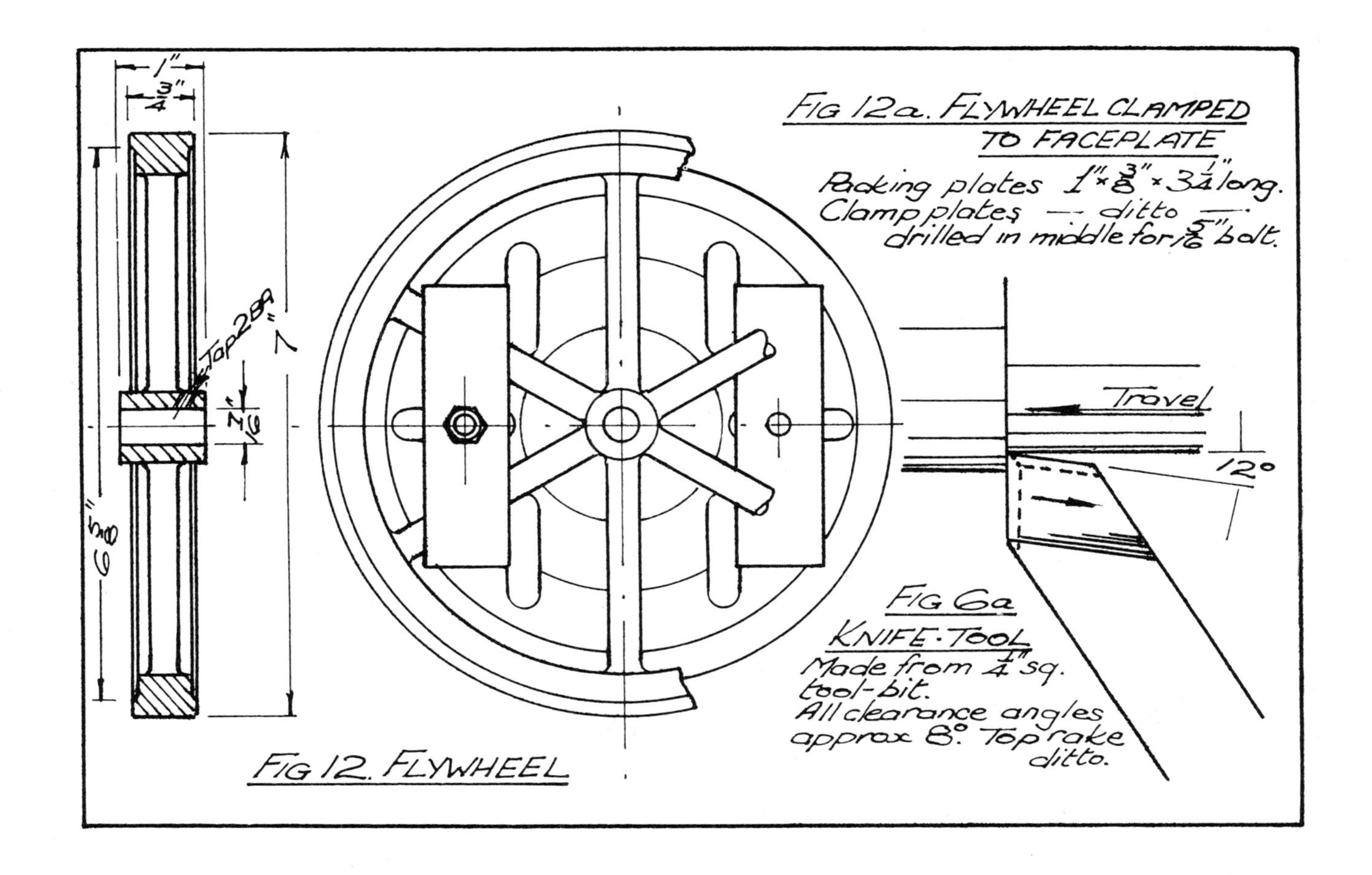

1"
3/4"
7"
6 5/8"
Tap 2BA
7/16"
FIG 12a. FLYWHEEL CLAMPED TO FACEPLATE
Packing plates 1" × 3/8" × 3 1/4" long.
Clamp plates — ditto —
drilled in middle for 5/16" bolt.
Travel
12°
FIG 6a
KNIFE-TOOL
Made from 1/4" sq. tool-bit.
All clearance angles approx 8°. Top rake ditto.
FIG 12. FLYWHEEL

A view from the rear showing main-bearings, crankshaft and flywheel.

extra weight in the flywheel helps to give a smooth running engine especially at low speeds. Increase the lathe speed now to the lowest in open gear and centre drill, open out and ream or bore to 7/16 inch for the crankshaft. Now reverse on the faceplate, set true and finish turn, leaving the boss 1 inch long and the rim $\frac{3}{4}$ inch wide. With a smooth flat file remove the sharp edges and the flywheel is complete apart from drilling and tapping for the 2BA grub screw.

TAPPING AND CLEARING SIZE DRILLS

May I draw your attention to this list of tapping and clearing drill sizes. The beginner may be surprised to see that, for example, he could choose any one of ten different drills as tapping size for a 5BA screw. In fact, the difference in diameter between the first and last of these listed is only $8\frac{1}{2}$ thousandths of an inch, that is just twice the thickness of the paper on which this little book is printed. Why the list? Well, due to metrication, number and letter size drills are being phased out and eventually Imperial sizes as well, but most of us will be using those we have for, I hope, a long time yet. If you are a beginner, choose a drill at the bottom of the list and you will have no difficulty in achieving successful tapping. If more experienced, work about the middle of the range and you will get a thread of fuller form. If you use a drill at the top of the list, then for goodness sake keep a steady hand. I have just paid over £1 for a high-speed steel 7BA tap!

Screw size	*Tapping size drill*	*Clearing size drill*
2BA	No. 26	3/16 inch
	No. 25	4·80 mm
	3·80 mm	No. 12
	No. 24	No. 11
	3·90 mm	4·90 mm
	No. 23	No. 10
	5/32 inch	
	No. 22	
	4·00 mm	
4BA	No. 34	No. 27
	2·85 mm	3·70 mm
	No. 33	No. 26
	2·90 mm	
	No. 32	
	2·95 mm	
	3·00 mm	
5BA	No. 40	No. 30
	2·50 mm	3·30 mm
	No. 39	3·40 mm
	2·55 mm	No. 29
	No. 38	
	2·60 mm	

Screw size	*Tapping size drill*	*Clearing size drill*
5BA	No. 37	
	2·65 mm	
	2·70 mm	
	No. 36	
6BA	No. 44	No. 34
	2·20 mm	2·85 mm
	2·25 mm	No. 33
	No. 43	2·90 mm
	2·30 mm	No. 32
	2·35 mm	
	No. 42	
7BA	No. 48	No. 39
	1·95 mm	2·55 mm
	5/64 inch	No. 38
	No. 47	2·60 mm
	2·00 mm	No. 37
	2·05 mm	2·65 mm
	No. 46	2·70 mm
	No. 45	No. 36

3. The Cylinder and all its parts

I begin to feel very embarrassed continually using superlatives when describing the quality of Stuart Turner castings, but when they are compared with some of those, both model and full-size, I have had to use the reason is very obvious. The beam engine cylinder we are using for "Victoria" is no exception, it is a beauty. Steam and exhaust ports and steam passages are cast in accurately and, certainly in my specimen, no work was necessary on them.

The cylinder should be cleaned all over with files. Not because it is strictly necessary, but I feel that anything we can do to save expensive power and help our pricey machinery and tools to last longer, is worth while. Apart from that, it is good practice for our filing skills and an excellent way to keep warm and exercise our muscles. No doubt you will deduce from this that the cold winter nights are upon us!

After cleaning, I found I could just grip the cylinder in the four-jaw chuck with the port-face outwards. The chuck is the 4½ inch light type (Burnerd) which, I imagine is about the smallest anyone will be using for this work. The cylinder located in the chuck jaw channel which acted like a vee block to help steady it, and the distance from the chuck face to the cylinder centre-line was assessed. This dimension, added to ⅞ inch gave the rule measurement at which facing of the port face had to stop. Just over 1/32 inch had to be removed, two very light cuts and the job was done. As this method of mounting the cylinder had been so successful, two jaws were slackened and the cylinder rotated 90° about its longitudinal axis, bringing the exhaust flange face outwards and this machined.

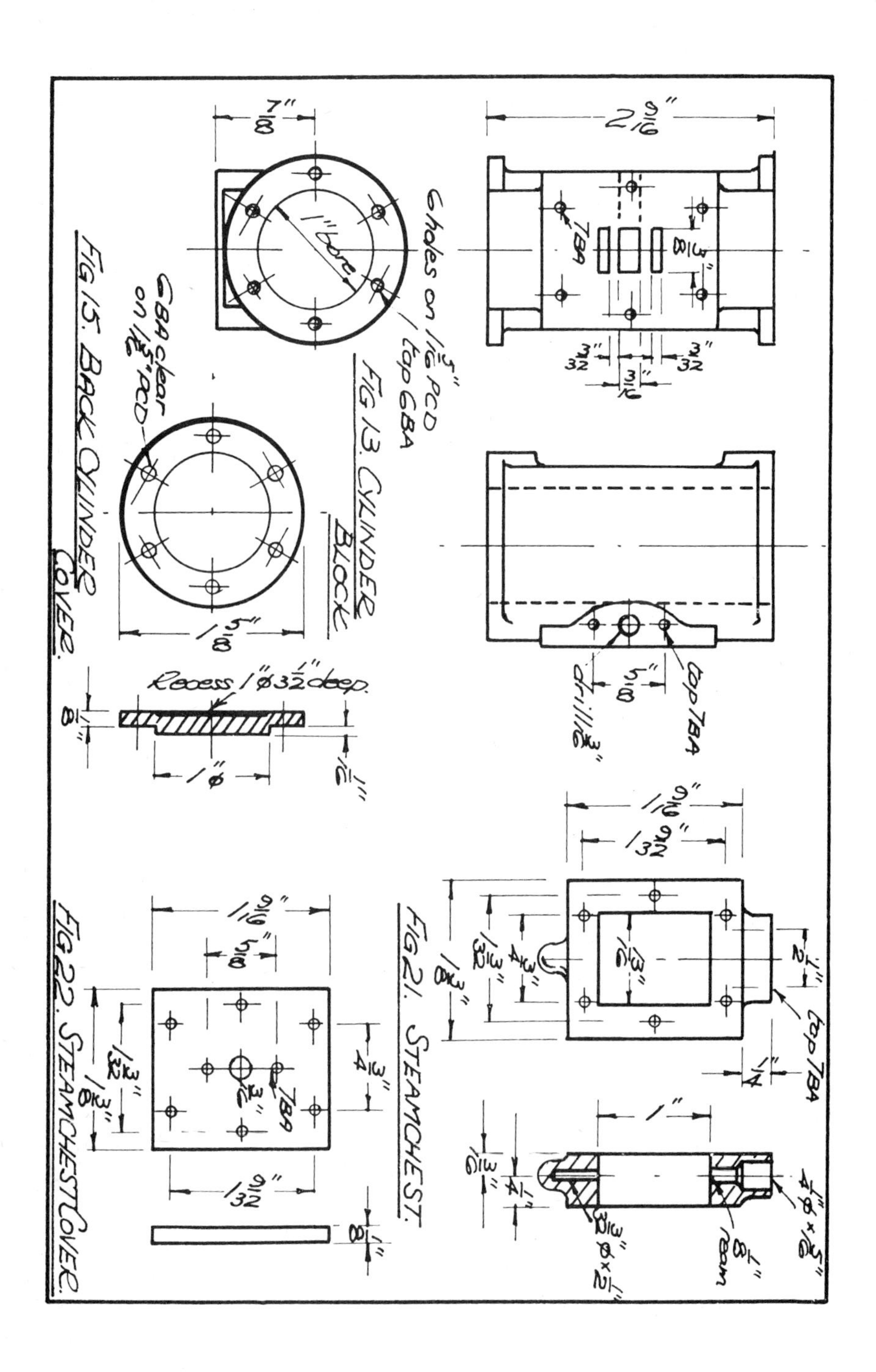

2 9/16"
7/8"
6 holes on 1 5/16" PCD / tap 6BA
1" bore
7BA
3/8"
3/32"
3/32"
3/16"
tap 7BA
drill 3/16"
5/8"
FIG 13. CYLINDER BLOCK
6BA clear on 1 5/16" PCD
1 5/8"
FIG 15. BACK CYLINDER COVER.
Recess 1"ø 1/32" deep.
1/8"
1/16"
1"ø
1 9/16"
1 9/32"
1 3/16"
3/4"
1 3/32"
1 3/8"
1/2"
tap 7BA
1/4"
FIG 21. STEAMCHEST.
1"
3/16"
1/4"
3/16"ø x 1/2"
1/8" ream
1/4"ø x 5/16"
1 9/16"
5/8"
3/4"
3/8"
7BA
1 3/32"
1 3/8"
1 9/32"
1/8"
FIG 22. STEAMCHEST COVER.

To bore the cylinder and face the ends, an angle plate bolted to the faceplate was used. The angle plate was made many years ago from a scrap of 3 inch angle and has been worth "a guinea an ounce" ever since. The cylinder was mounted, with a protective piece of paper, on its port-face, so that the outside of the flanges ran true. Longitudinal accuracy was taken care of by checking from the faceplate with a trysquare.

The cored hole through the cylinder is $\frac{7}{8}$ inch diameter which allows a sufficiency of metal to permit an accurate bore, without the tedium of a multitude of long slow "boring" operations on a light lathe. Mark the end of the cylinder where bore and flange were machined at the same setting, this will be the gland or "front" end when the engine is assembled.

I suggest that we now screw the four-jaw chuck back on the lathe, because many of the remaining components require this facility and it will save some time if we group these operations together.

First, we might start on the steam chest, Fig. 21. After cleaning with a file, especially the 1 inch by 13/16 inch interior, we can face all the necessary surfaces to size. The important thing here is to notice the difference in thickness ($\frac{1}{4}$ inch and 3/16 inch) on either side of the centre-line. I suggest that you keep the $\frac{1}{4}$ inch dimension fairly full, as any reduction here will limit the clearances available when fitting the slide-valve. The final operation should be to set the gland to run true, face, centre drill and drill 3/32 inch continuing on into the opposite end of the chest. Then open out with a No. 31 drill, through the gland boss only, and ream $\frac{1}{8}$ inch. Finally open out with a $\frac{1}{4}$ inch drill to a depth of 5/16 inch. Mark out and drill the six 7BA fixing holes and, while you are about it, use these holes to spot through on to the port face of the cylinder, drill tapping size and tap 7BA. Start the taps in the drilling machine so that the threads will be square with the surface.

The steamchest cover looks awkward to hold for machining, but is so accurately cast to size that it can be completed by careful filing. Rub the surfaces on a flat, smooth file and finish on emery cloth. If you feel competent, by all means face in the lathe, holding in the four-jaw chuck. When finished to size, clamp to the upper surface of the steam chest and drill through from the six 7BA clearance holes in the latter. Finally, in the middle of the cover, drill a 3/16 inch hole for steam admission.

While the four-jaw chuck is mounted on the lathe, let's have a session of elliptical gland and flange machining. These, luckily, are supplied as neat little gunmetal castings, and I suggest tackling them all together. Clean them up by filing, (Fig. 17, 31, 23 and 24), but do not, for the present, reduce the elliptical part to finished size. The "modus operandi" is, more or less, the same for each of these pieces, so one description will suffice.

Gripping by the elliptical flange, mount the casting in the chuck with the boss outwards and running as true as possible. Hold so that sufficient of the flange is protruding to allow a light facing cut to be taken over it. Face the end of the boss and, if you feel at all uncertain, centre drill the boss so that it can be supported by the tailstock. Turn the boss to diameter and skim the flange at the same time. Now drill the boss and in the case of the glands ream 3/16 inch and $\frac{1}{8}$ inch respectively, while the steam flanges are tapped, 3/16 inch by 40 tpi for the inlet and $\frac{1}{4}$ inch by 40 tpi for the exhaust. Later the glands can be set in their respective positions and the ellipses filed and polished to blend nicely with the gland bosses. The steam flanges should be filed until they give a shape which is pleasing to the eye.

Finally, mark out and drill for the fixing bolts and studs.

We might now turn our attention to the piston assembly, and so that we can finish turn the piston on its rod we will make this first, Fig. 19. It is a length of 3/16 inch diameter stainless-steel and because it has been precision ground, we should be very careful not to mark or damage its surface. Nothing causes leaky glands quicker than packing which is being abraded by piston or valve rods whose surface is marked by having been gripped, without protection, in chuck or vice.

Make a small split bush, Fig. 19a, from a scrap of brass or aluminium rod, mark it to show the location of No. 1 chuck jaw and split it with a junior hacksaw. A similar bush, made from a piece of square bar, may be necessary if the 2BA thread needs completing in the bench vice.

Hold the 3/16 inch rod by the bush in the self-centring chuck, face the end and turn a $\frac{1}{2}$ inch length to $\frac{1}{8}$ inch diameter, thread 3/16 inch of this to 5BA. Reverse in the chucking bush and face down to a length of 3-13/32 inch from the shoulder and, with the diestock supported by the tailstock barrel, cut a 2BA thread, leaving a plain portion in the middle of 3-5/32 inch long. At a later date, the length of 2BA thread will need facing down so that the piston, piston rod and piston-rod end assembly are to the overall dimensions shown in the drawing of the assembled components.

The piston, Fig. 18, is a gunmetal casting with cast-on chucking spigot. Clean up with a file, then mount by the piston in the self-centring chuck with the spigot outwards. Removing the minimum of metal, turn the spigot true and face the side of the piston. Reverse in the chuck, gripping by the spigot, face the piston to $\frac{3}{8}$ inch length, and lightly centre drill. With the tailstock to help keep things steady, turn the outside of the piston to a good 1/32 inch oversize; then with a 1/16 inch wide parting tool, machine the packing groove. Drill the piston $\frac{1}{8}$ inch and with a tiny boring tool, recess the face of the piston to take the 5BA nut.

Reverse the piston in the chuck and face off the remains of the chucking spigot. Clear any burr from the hole in the piston then, with suitable protection, grip the piston rod by the 2BA end in the tailstock chuck and gently press the 5BA end through the hole in the piston. Remove the assembly and fit the 5BA nut. Remount by the rod (and bush) in the lathe chuck and face down the 5BA thread to give a neat finish. Then, using the cylinder as a gauge and with a really sharp tool, carefully skim the piston, removing only a gnat's whisker at a time, until it slides without slop into the cylinder. Remove piston and rod from the lathe and put it somewhere, where there is absolutely no risk of it being damaged!

Now to the cylinder covers. The front cover, Fig. 14, is a bit awkward, as it has no chucking spigot. The sequence of operations is as follows. Grip in the chuck by the locating spigot, gland boss outwards. Face the boss to 5/16 inch long. Centre drill and steady with the tailstock. Turn the periphery to $1\frac{5}{8}$ inch diameter and face an annular rim on the cover for the heads of the fixing screws to bed on, then with

Cylinder being bored, mounted on angleplate bolted to faceplate. Lathe changewheels used as balance weights

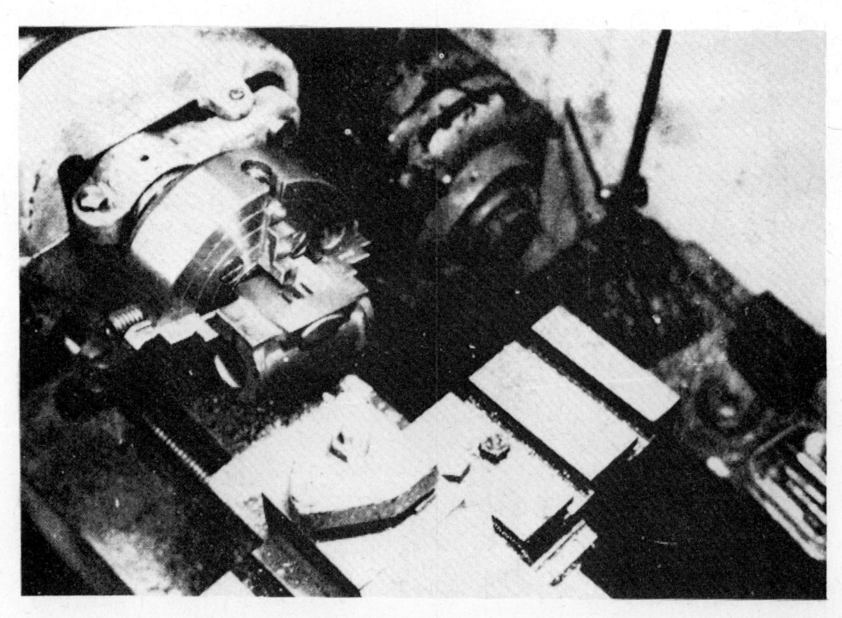

Machining the exhaust flange facing

The portface just completed

Machining surfaces of steam-chest

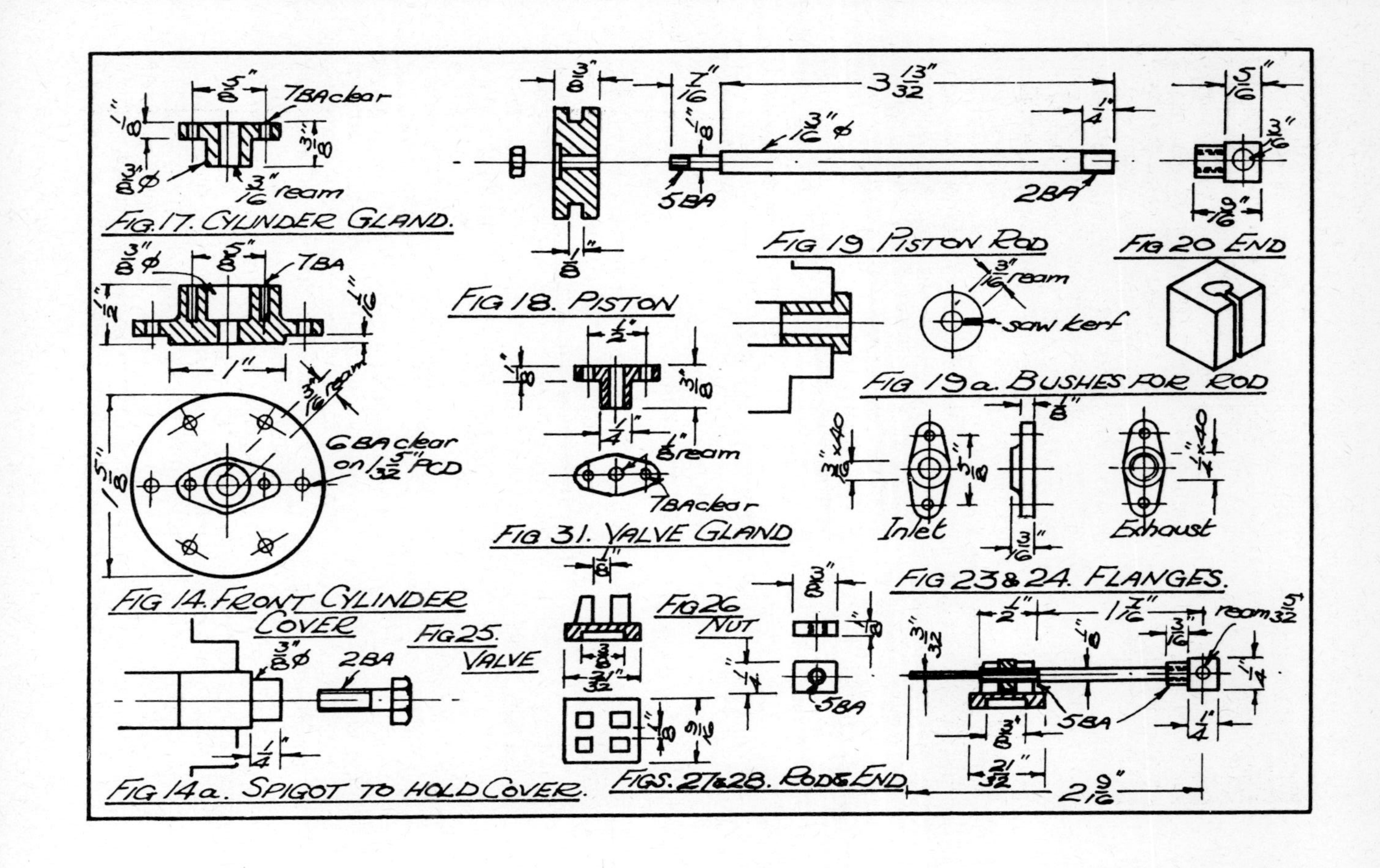

7BA clear
3/16" ream
FIG. 17. CYLINDER GLAND.
7BA
3/16 ream
6BA clear on 1 5/32" PCD
FIG 14. FRONT CYLINDER COVER
5BA
2BA
FIG 19 PISTON ROD
FIG 20 END
FIG 18. PISTON
1/8 ream
7BA clear
FIG 31. VALVE GLAND
3/16" ream
saw kerf
FIG 19a. BUSHES FOR ROD
Inlet
Exhaust
FIG 23 & 24. FLANGES.
FIG 25. VALVE
FIG 26 NUT
5BA
ream 5/32
FIG 14a. SPIGOT TO HOLD COVER.
FIGS. 27 & 28. RODS END

the tailstock chuck, drill and ream the boss 3/16 inch. Open out to $\frac{3}{8}$ inch diameter for a depth of 5/16 inch, and check that the gunmetal gland fits easily.

To turn the 1 inch diameter locating spigot, we will require something to hold the cover so that the rest of our turning will be concentric with the hole in the gland. This chucking mandrel is shown in Fig. 14a. Do not remove this accessory from the lathe until you have completely finished with it. With the cover mounted on it, face to the thickness shown on the drawing and aim for a really close fit of the cover in the cylinder bore. Make sure you are checking with the right end of the cylinder. The little bit remaining under the head of the 2BA bolt can be faced off with the cover held by its edge in the chuck.

The back cover, Fig. 15, is a simpler version of the previous procedure and requires no detailing.

In each case, while the cover is in the chuck, mark a circle of 1-5/16 inch pitch circle diameter with a vee-tool and on it locate, by centre punch, the position of the 6BA clearance holes. Drill these, then use the covers to locate the holes on the cylinder ends. Drill and tap these holes in the cylinder 6BA. The two lower holes, in the covers, which come under the cylinder brackets, may be countersunk, if this is considered more suitable, i.e. if the screws come near the edge of the cylinder bracket.

Use the cylinder gland to spot for the two 7BA tapped holes in the gland boss. Drill and tap these holes and fit two studs. When fitting studs, remember that the end with the short thread is the 'metal end,' i.e. is screwed into the work. The end with the longer thread takes the nut.

Likewise, the 7BA holes for the exhaust flange can be completed if not already done.

The slide valve and its bits and pieces might well form our next bit of work. As we have the three-jaw chuck set up, let's make the valve spindle first, Fig. 27. This is from $\frac{1}{8}$ inch diameter stainless steel. In order to achieve concentricity and avoid marking the surface of the spindle, it would be worth while making a little bush, as previously described for the piston rod. With things organised, turn down to a bare 3/32 inch diameter for a length of $\frac{5}{8}$ inch, aiming for a fine finish and smoothly dome the end with a fine file. Draw the work out further from the chucked bush and thread the next $\frac{1}{2}$ inch length, 5BA. Stainless steel can be a little awkward at times, so use plenty of cutting oil. Reverse the spindle in the chuck, face to an overall length of 2-9/16 inch, and thread this end 5BA for a length of 5/16 inch. Finally face down to a final length of 2-7/16 inch, after you have checked the threads.

A session with the four-jaw chuck is now called for; first we will make the valve nut, Fig. 26. This is a very simple component, but it is crucial that the 5BA tapped hole be accurately square with the nut surfaces. A tiny gunmetal casting may be supplied for this. In my case I made it from a short piece of $\frac{1}{4}$ inch by $\frac{1}{8}$ inch brass bar. Nevertheless, I set it up in the four-jaw chuck to ensure that the tapped hole was square, and unless you are absolutely sure of your drilling machine, I suggest you do likewise.

Now grip the slide valve, Fig. 25, in the chuck, face outwards. Lightly skim the surface taking off only sufficient to obtain a good finish.

Chucking spigot on which cylinder cover will be mounted for final machining

Machining front cylinder cover

If we take off too much, it may necessitate having to deepen the cavity which is a bit of a nuisance. Machine the sides of the valve by lightly facing, or if only a tiny amount needs to be removed, by filing, until its overall size is, as shown, 21/32 inch long by 9/16 inch wide, with the cavity $\frac{3}{8}$ inch by $\frac{3}{8}$ inch centrally situated.

Clean up the back of the valve with a file and mark the positions of the two $\frac{1}{8}$ inch wide slots. Carefully hacksaw the bulk of the metal out. With careful sawing, you will be surprised how little filing will be needed to clear out these slots. In spite of all the fancy machine tools that become available, I still think that the ability to saw so accurately that you can skim a scribed line is one of the most useful skills that a craftsman can develop. If a $\frac{1}{8}$ inch end mill or slot drill is available, you can finish to size, after sawing and filing, with the valve clamped to the vertical slide, or, as in my case, under the tool-post.

With the valve assembly mounted in the steam-chest, there must be clearance under the edge of the valve nut, so that the valve is able to lift slightly off the cylinder portface. In all other cases the nut must be a close sliding fit in the valve slots.

The rod ends Figs. 20 and 28, are straightforward examples of four-jaw chuck work, but like most apparently simple components, they have got to be made just right. In this case, the threaded hole to take the rod must be square with the cross-hole which takes the wrist pin. One description will suffice for both components. Face each end of the square mild steel bar to finish just over final length. Now set up cross-wise in the chuck and centre drill for the cross-hole, drill and ream as necessary. Mount longitudinally in the chuck and set to run true, centre drill, drill and tap as required, then turn down to finish as shown.

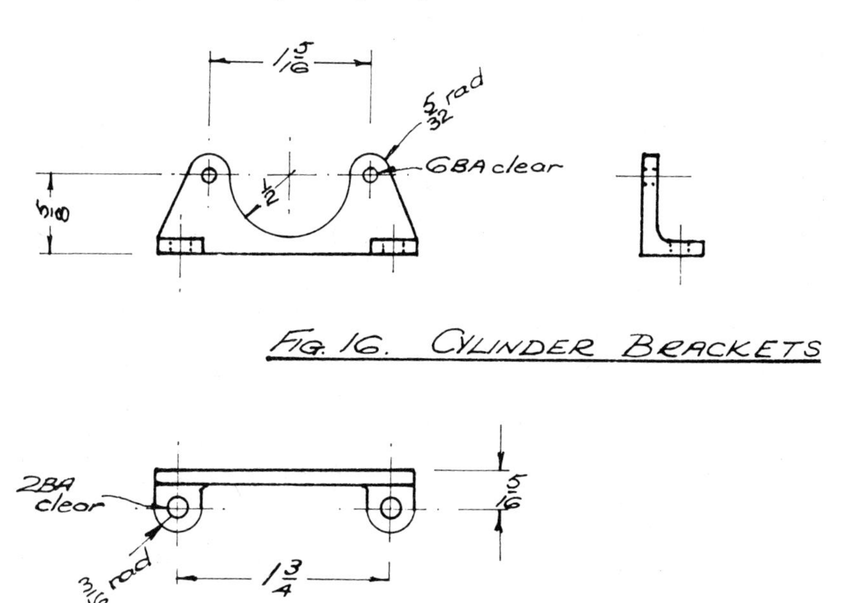

FIG. 16. CYLINDER BRACKETS

Very lightly chamfer the end. Reverse in the chuck and face down to the correct length.

The cylinder supporting brackets, Fig. 16, are made from 1 inch by 1 inch bright drawn mild steel or brass angle. Careful marking out, drilling and filing, with, of course our friends the filing buttons, will soon provide a substantial pair of brackets. Fit them to the assembled cylinders and check that they sit flat on the soleplate facings.

That completes the bits and pieces of the cylinder and we can now assemble with a smear of oil and congratulate ourselves on the accuracy of fit which is evident in the way the piston and valve slide to and fro. Do not, for the present, fit any gland or piston packing; we will leave that until the final assembly.

4. Connecting rod, crosshead and guides, valve operating gear

The connecting rod, Fig. 7, forms the opener for this group of components. With this piece out of the way, we are well on the road to seeing the "Victoria" project near completion.

A length of bright mild steel of $\frac{5}{8}$ inch by $\frac{1}{2}$ inch section is supplied for this and the first job is to file the ends square, mark out and centre drill them. This may be done with the bar held in the lathe four-jaw chuck or mounted upright in the drilling machine. Use the smallest size of centre drill.

These drills are expensive and, if broken, difficult to resharpen; and although, when broken, they make useful boring cutters, etc., it is an expensive way of building up a stock of high speed steel! Hence, if I am centring long bars in the drilling machine, I usually carefully centre punch then drill with a short stubby drill, of about the right diameter and finish off with the correct centre drill.

Now mark out accurately the position of the three holes needed in this component, Fig. 7a. These are the big-end—to be reamed $\frac{3}{8}$ inch for a bronze bush, the small-end—reamed 3/16 inch for the wrist-pin. and the 5/16 inch drilled hole forming the basis of the shaped small end. This latter, although only drilled, must receive as much, if not more care than the other two because if it does not finish up accurately in the centre of the 9/16 inch wide face it will spoil the whole appearance of the rod.

The first and second holes should at the moment be drilled $\frac{1}{8}$ inch or whatever size suits your filing buttons.

May I just say a word about these filing buttons. Up to quite recently I have made them with a $\frac{1}{8}$ inch diameter pip and hardened them fully—heat to cherry red and quench in water. But lately I have been plagued with the pip breaking off, a brittle fracture at the internal corner of the pip. I have therefore started to make them by drilling through $\frac{1}{8}$ inch diameter and clamping them to the work with $\frac{1}{8}$ inch screws and nuts. There appear to be a number of advantages; they are easier to make, just drilling and parting off; they take more kindly to hardening; and an extra bonus, they can be used as extremely useful drilling jigs using a $\frac{1}{8}$ inch pilot drill. Fig. 7b shows the idea.

Turning the connecting rod

Before turning the rod, I suggest that you saw and file an $\frac{1}{8}$ inch wide strip from each side, leaving the rod $\frac{3}{8}$ inch by $\frac{1}{2}$ inch section, apart from the portion at the crosshead end. If however, sawing is a skill which you have not yet mastered by all means leave it and remove the metal by turning. The final result will be the same.

Now mount the bar in the lathe, either between centres or with one end held in the four-jaw chuck and the other supported by the tailstock centre. With a round nosed tool turn the required portion, about $4\frac{3}{4}$ inches, to 11/32 inch diameter and parallel. The next step is to turn a taper at each end of this cylindrical portion to get the bellied effect. If your top-slide has a travel of at least 2 inches you can use this. Slacken its clamping nut and rotate it so that on a length of about $2\frac{1}{4}$ inches the inclination is 1/32 inch. This is best checked using a trysquare and steel rule, and preferably have the left hand end of the top-slide 1/32 inch nearer the work. This will allow you to work towards the chuck. Now using the same round nosed tool, turn the taper at the chuck end of the rod. Let the last cut be no more than a couple of "thou" deep and use plenty of cutting oil. Don't go below $\frac{1}{4}$ inch diameter at the narrow end. Reverse the work in the lathe and reset to run true. Repeat the process on the other end. Finally turn the centre portion parallel to 5/16 inch diameter and blend in and polish with some fine emery cloth.

Remove the rod from the lathe and reduce the big-end to 9/32 inch thick making sure to check with the turned part to see that the overall appearance is symmetrical. Using filing buttons, round the end to $\frac{1}{4}$ inch radius then carefully set up on the drilling machine and open out to 23/64 inch and ream $\frac{3}{8}$ inch.

Reduce the $\frac{1}{2}$ inch thickness to a full $\frac{3}{8}$ inch at the crosshead end and with filing buttons radius to $\frac{3}{8}$ inch diameter, file down the remainder to $\frac{1}{4}$ inch thick to blend in nicely with the turned portion. Open out and ream the 3/16 inch hole for the wrist pin. Saw out and file the centre of the yoke and with a smooth file, file all the flat surfaces to give a good finish.

While it is in order to finish bright steel surfaces with fine emery cloth, be careful not to round off the corners too much. It is the way in which light is reflected from bright steel surfaces at various angles that is largely responsible for the attractive appearance of a well finished stationary steam engine model. If the corners are rounded much of the crispness and sparkle is lost.

From a short piece of bronze or gunmetal make the bush for the big-end. After pressing in, pass a $\frac{1}{4}$ inch reamer through to open out to crankpin size. Drill and countersink an oil hole no bigger than 1/16 inch diameter through rod and bush. Finally drill a clearance hole and fit a 7BA bolt and nut as shown. This is purely for appearance, but the rod looks rather bare without it.

The crossheads, Fig. 38, are made from hard drawn brass. Face the blocks in the lathe four-jaw chuck to an overall size of $\frac{3}{4}$ inch by $\frac{5}{8}$ inch by $\frac{1}{2}$ inch. With suitable packing, clamp them under the tool post and mill the steps, leaving the centre $\frac{3}{8}$ inch wide. Then hold in the four-jaw chuck with the opposite face outwards, centre drill, drill and ream to 3/16 inch and reduce the outside to $\frac{3}{8}$ inch diameter.

Drilling and parting off the guide bar supports

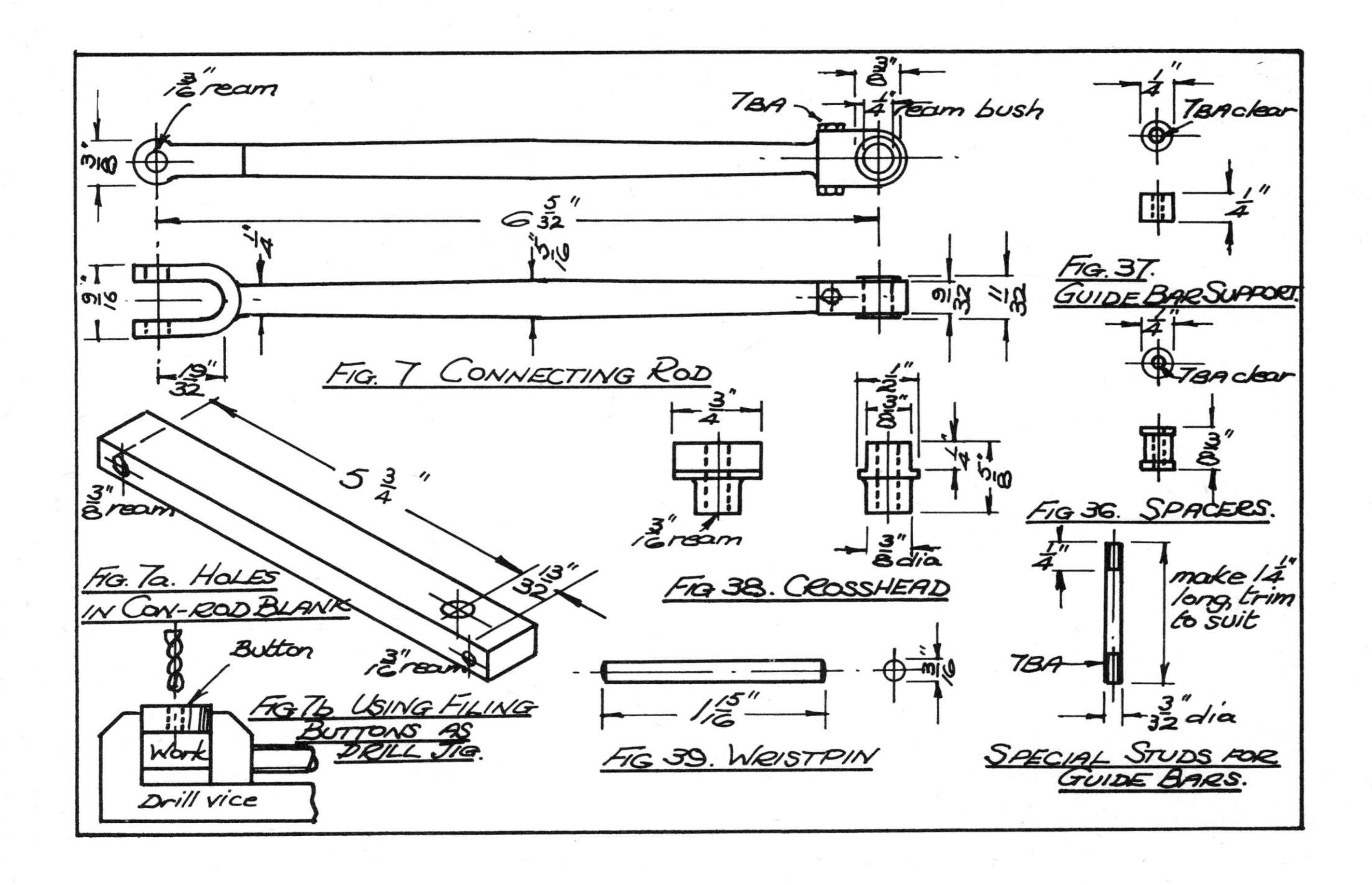

FIG. 7 CONNECTING ROD

FIG. 7a. HOLES IN CON-ROD BLANK

FIG. 7b USING FILING BUTTONS AS DRILL JIG.

FIG 38. CROSSHEAD

FIG 39. WRISTPIN

FIG. 37. GUIDE BAR SUPPORT.

FIG 36. SPACERS.

SPECIAL STUDS FOR GUIDE BARS.

If we make the wristpin now, we can assemble the crossheads to the connecting rod. It is a very simple job, Fig. 39, but for a good appearance, aim for nicely domed ends.

With the three-jaw chuck on the lathe, we can machine the guide bar supports. These again are very simple turning exercises, Fig. 37. Two are plain and finished to $\frac{1}{4}$ inch long, while the upper two spacers, Fig. 36, have their centres reduced in diameter and are nominally $\frac{3}{8}$ inch long. I say nominally because they relate to the width of the crossheads which must slide easily, but without slop, between the guide bars. For the moment, it would be sensible to leave these spacers fractionally over length.

The lower guide bars, Fig. 35, are simple filing and drilling jobs from $\frac{1}{4}$ inch by 3/16 bright mild steel. A little more work is entailed in the upper bars, Fig. 34, and for a neat finish it is worth milling the little recesses at each end. Facing the ends of these bars is conveniently done in the lathe. A tiny oil hole drilled in the top bars will take care of lubrication, or they may be tapped $\frac{1}{8}$ inch by 40 tpi for the smallest size of Stuart Turner oil cups (part No. 183/1) which would look rather nice here.

To assemble the guide bars we require four extra long 7BA studs, see sketch. These are not available from Stuart Turner, but may be easily made from a length of 3/32 inch diameter mild steel rod. Make them $1\frac{1}{4}$ inches long with $\frac{1}{4}$ inch of 7BA thread at each end and trim them to size as required. It is, needless to say, important that threading be done in the lathe to ensure that these studs stand perpendicular to the base of the engine. During assembly, the guide bar spacers should be carefully faced off so that the crossheads are a close sliding fit between the guide bars.

As we can no longer get a small boy to open and close a steamcock to admit and release steam to and from the cylinder, for a few coppers a week, we must replace him with a valve operating gear! Start with the eccentric strap, Fig. 8. This is a gunmetal casting which after cleaning up with a file requires the holes for the 7BA clamping bolts to be drilled tapping size. Now carefully saw the strap in half and face the sawn surfaces by rubbing on a smooth flat file. The 7BA holes may now be tapped or opened out as required in the two halves. Now bolt them together; use a couple of spare 7BA bolts, and set up the strap in the four-jaw chuck with the bore running as truly as possible. Bore to 1-1/16 inch diameter and recess. Aim for a really fine finish. Face the outside of the strap removing about half of the excess thickness of metal but making sure that eventually the clamping bolts will finish in the centre of the strap width. Reverse in the four-jaw chuck, make sure that the faced side will run true by setting up with some parallel packing between it and the chuck face and machine the second side to the finished width of $\frac{1}{4}$ inch. Deburr and drill a tiny oil hole. Finish the strap on which the rod is bolted by filing or milling in the lathe.

The eccentric sheave, Fig. 11, is turned from a short length of bright mild steel. Face then use the strap as a gauge for turning the 1-1/16 inch diameter and the raised rib. When this part has been completed, reverse the sheave in the chuck and face the opposite side. Remove from the

Boring the eccentric strap

lathe and with steel rule and scriber, draw a line diametrically across the faced surface. From the centre of the sheave, which will be evident from the turning marks, measure out 9/64 inch and lightly centre punch. Reset in the four-jaw chuck and set the centre punch mark to run true. Centre drill, open out and ream 7/16 inch, then turn the outside down to 11/16 inch diameter. Face to the overall lengths shown in the drawing. Drill and tap for the 5BA grub screw.

The eccentric rod, Fig. 9, and fork, Fig. 10, is a simple bit of careful bench work, but keep the rod as long as possible for the time being and don't drill it for bolting to the eccentric strap until you check the actual centre distances on final assembly. To find a rod is fractionally too long or short after a lot of careful work has been carried out on it is annoying and things never seem to work out exactly according to the drawings no matter how carefully we think we have followed them. In fact, if you look at the photographs of the prototype of "Victoria" which I built, you will see that the eccentric rod is cranked, due to a modification carried out in the early stages which was not such a good idea as I had originally imagined!

Because the valve chest on "Victoria" is above the cylinder and therefore not in line with the eccentric rod, the drive to the valve spindle must pass through a simple lever system. If we start with the so-called "bridge", Fig. 40, which carried the levers, the other parts can follow. This is made from a 2 inch length of $\frac{1}{4}$ inch by 7/16 or $\frac{1}{2}$ inch bright mild steel. Mount in the four-jaw chuck and face each end to $1\frac{7}{8}$ inch overall length, then mark out the position of the 5/32 inch hole, set this to run

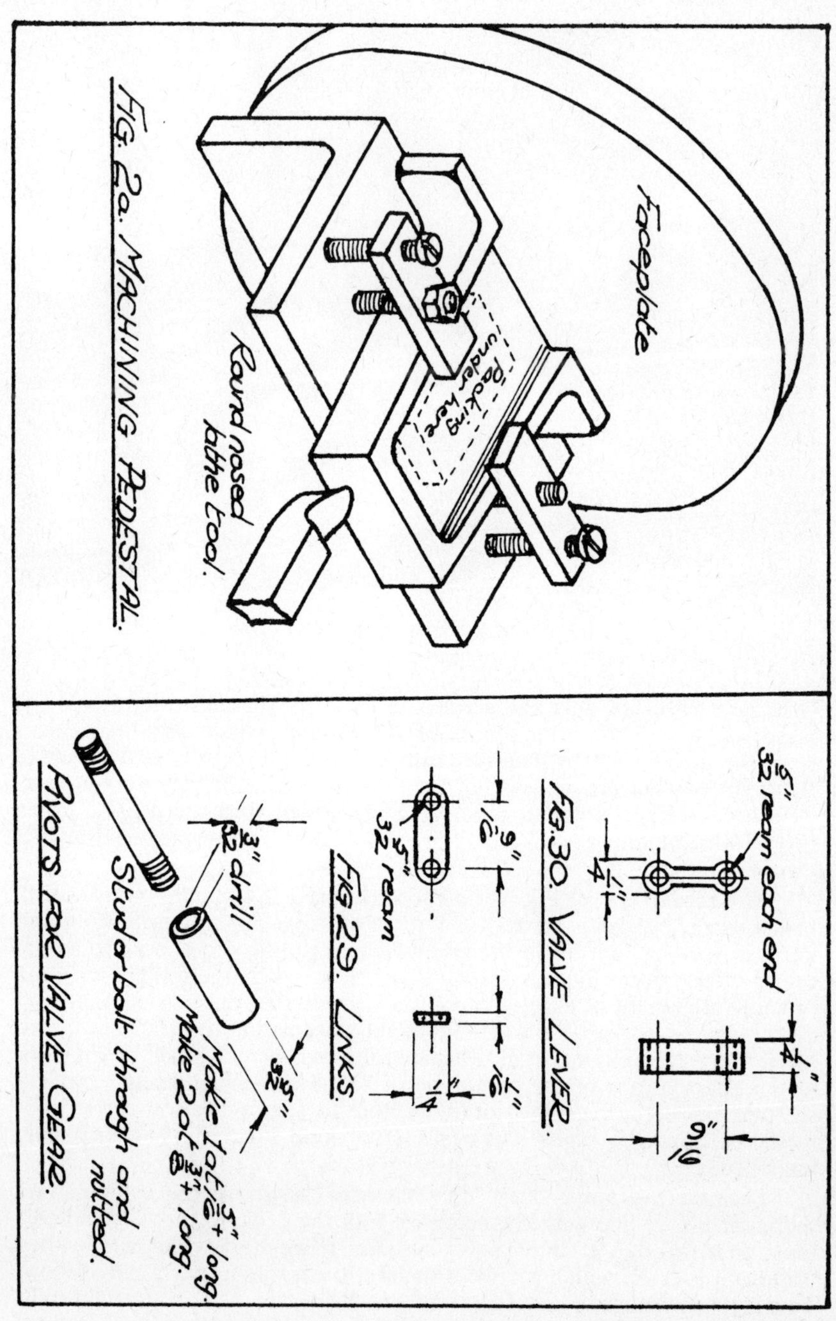

The Stuart Turner drawings show the Valve Gear Pivots or Link Pins (Part No. 41) *as being turned from* 5/32 *inch diameter rod.*

true and centre drill, open out and ream this hole right through the length of the steel. You may not have a 5/32 inch reamer—I didn't—so drill just below this size and finish with a new 5/32 inch drill. Mark out and file or mill the recesses shown, round the top edge and drill a couple of tiny oil holes. Finally drill the two 7BA clearance holes. The rod which passes through the bridge, Fig. 33, is a simple turning job on 5/32 inch diameter steel bar, preferably silver steel as the bridge piece in which it rotates will not be bushed.

Little need be said about the levers, Figs. 30 and 32, and the links, Fig. 29, except to aim at a good finish as all these bits are particularly visible. In the levers, it is important that the holes be perpendicular with the surfaces and parallel with each other so that they do not bind. The pivots are made as detailed, from 5/32 inch silver steel rod, drilled through 3/32 inch, with studs threaded 7BA and fitted with nuts. I find that this is the easiest way to make such pivot pins for valve gears, etc., the outer bush should be made from silver steel, and either left soft or hardened and tempered if for arduous duties.

If you check you will now find that we have completed all the constructional work required in building this stationary engine. You can congratulate yourself on having successfully made over forty individual parts in cast-iron, steel, brass and gunmetal. If this has been your first venture into steam engine building you can feel justly proud of your achievement and gain confidence from the knowledge that you now have the skill and experience to tackle any similar model (or real) engineering project within the limits of your workshop equipment.

This pigeon's eye view helps to show the valve-operating gear.

In this chapter there are two points which require some explanation. The first concerns a note which now appears on the detail drawing for the "Victoria" engine supplied by Stuart Turner Ltd., and states:

"If ports vary from dimensions given, alter valves to suit."

The details relating to this note appear in the diagram contained in this chapter, which shows to an enlarged scale and fully dimensioned, the section through the steam and exhaust ports and the slide valve.

The important features are first, that the exhaust cavity in the valve should truly match, line for line, to the inner edges of the steam ports on the port face; similarly, the overall length of the slide valve, quoted on the drawing as 21/32 inch, should be such that it has an overall length that is 3/32 inch greater than the overall length of the steam and exhaust ports, measured along the port face.

Arrangement of guide-bars, cross-heads and motionwork.

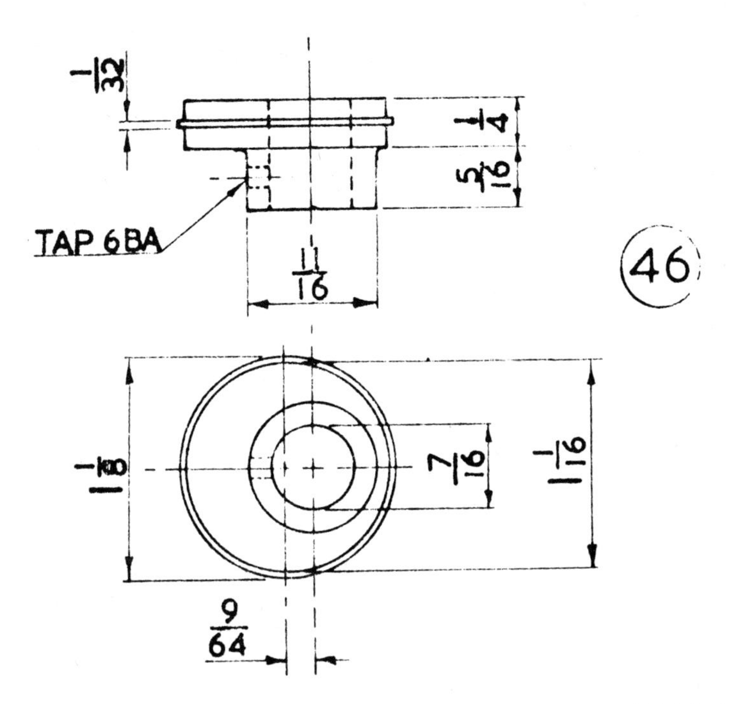

The eccentric sheave (*Part No.* 46) *showing the new reduced throw of* 9/64 *inch, replaces part Fig. 11.*

The amount by which the slide valve overlaps the steam ports should be equal at each end. This is called the lap of the valve.

The second point is the difference in throw of the eccentric sheave. This is shown in earlier drawings and in the first edition of this book, as 11/64 inch, in later drawings and in the present edition, it is shown as 9/64 inch. Again, the reason for this alteration is seen on the previously mentioned diagram.

With a throw of 11/64 inch the valve would open the steam port and continue to travel beyond the fully open steam port for a further 1/32 inch. This results in a wasteful use of steam, and so by reducing the valve travel to 9/64 inch the port is opened fully but for a shorter period of time thus allowing steam to be used more economically. In fact, the eccentric sheave throw could be reduced to ⅛ inch allowing a very small degree of expansive working. I hasten to point out that it is only within the last year or so, that this alteration in the throw of the eccentric sheave has been made. All engines previously built and having the longer valve travel will work equally well and in every way to their builder's satisfaction.

A separate drawing showing the correctly dimensioned eccentric sheave is included in this chapter.

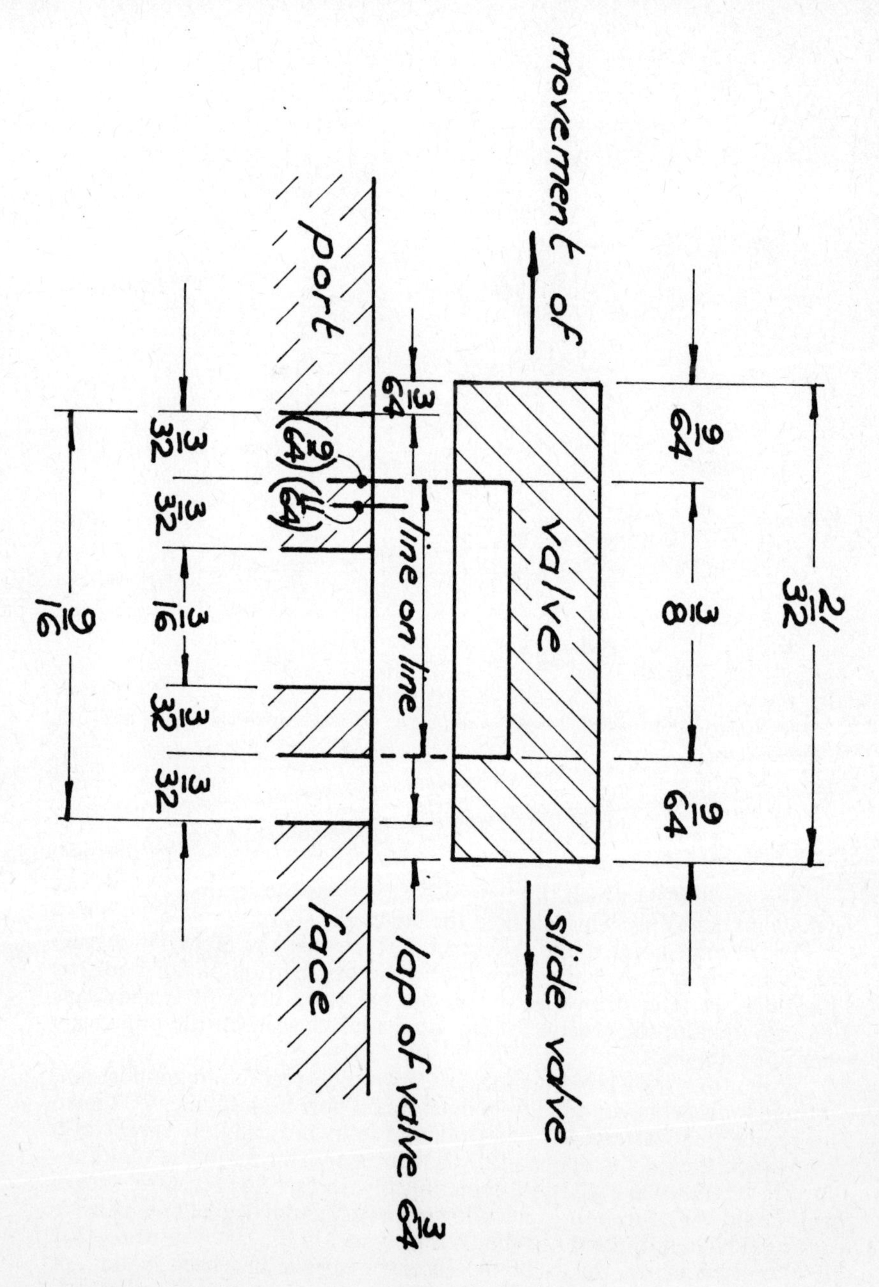

This diagram illustrates the relationship between the steam and exhaust ports and the slide valve.
The two lines marked (11/64) *and* (9/64) *show the limit of travel of the slide valve when actuated by eccentric sheaves of different throws.*

5. A woodworking session, erecting the engine, the final bits and pieces

No doubt as you worked through this project, you have been slowly erecting the engine. Before final completion can take place it is obvious that some kind of base is required to support the engine and the outrigger type of pedestal bearing and to raise the soleplate to the level of the pedestal. On the original, the engine base was made of timber of about 10 inches by 5 inches section, with the biggest dovetail joints I have ever seen at the corners. Needless to say, these details of construction only became evident after about an inch of a greasy filth had been removed. A centre member was mortised through the side pieces and tied by a long nutted stud. The drawing of the base details show what is required. Make it from hardwood, beech or mahogany would look nice, or any close-grained timber. Finish with a clear polish or varnish. Notice that the finished section of the timber is 2 inches by 1 inch, hence it must be of larger section before planing. In other words, don't make the mistake, if you have to buy this wood, of asking for a length of "two by one!"

A "floor" to carry engine and outer bearing is needed and as the original floor on which friend John and I sweated, heaved and strained was of boarding, I represented this on an odd piece of $\frac{1}{2}$ inch thick birch plywood by drawing lines 1 inch apart with a black ball-point pen and varnishing over. This has made the engine easy to handle on its occasional trips up to Henley-on-Thames. The same idea might conveniently be used if you envisage the engine as the main feature of a scale representation of the engine room of a small back-street workshop, complete with boiler, shafting, pulleys; all perhaps driving a small dynamo to light this steam powered display.

As hinted at the beginning of this chapter, your engine has probably been erected as work progressed. If so, give it a thorough check over and make a note of any details which need dealing with before the final painting and assembly takes place. Such things as studs and bolts which may need trimming to length, deburring of sharp edges, drawfiling vice marks out of bright pieces and mentally determining to remember to use soft vice clamps in future! Check that the oil holes have been drilled including two in the bridge (40), that the flywheel rotates without wobble, that the steam inlet hole has been drilled in the steamchest cover, and the exhaust hole in the cylinder.

With all these details taken care of, now is the time to dismantle the engine and give some thought to painting it. I cannot tell you the colours of the original, except to say that where there were layers of greasy dirt, the final surface colour appeared to be black, although this could have been any dark colour initially. Where the surfaces had not been protected by oil, old man rust had been having a sumptuous feast. When I had finished with the full-size engine its main colour was dark green with moving parts picked out in red, shafts in polished steel and glands polished bronze. I was sorry to part with it.

My particular model, which is before me as I write, has the inside of the soleplate a light blue-grey; the main colour for soleplate, cylinder and steamchest, flywheel and pedestal, dark green; crankweb, red;

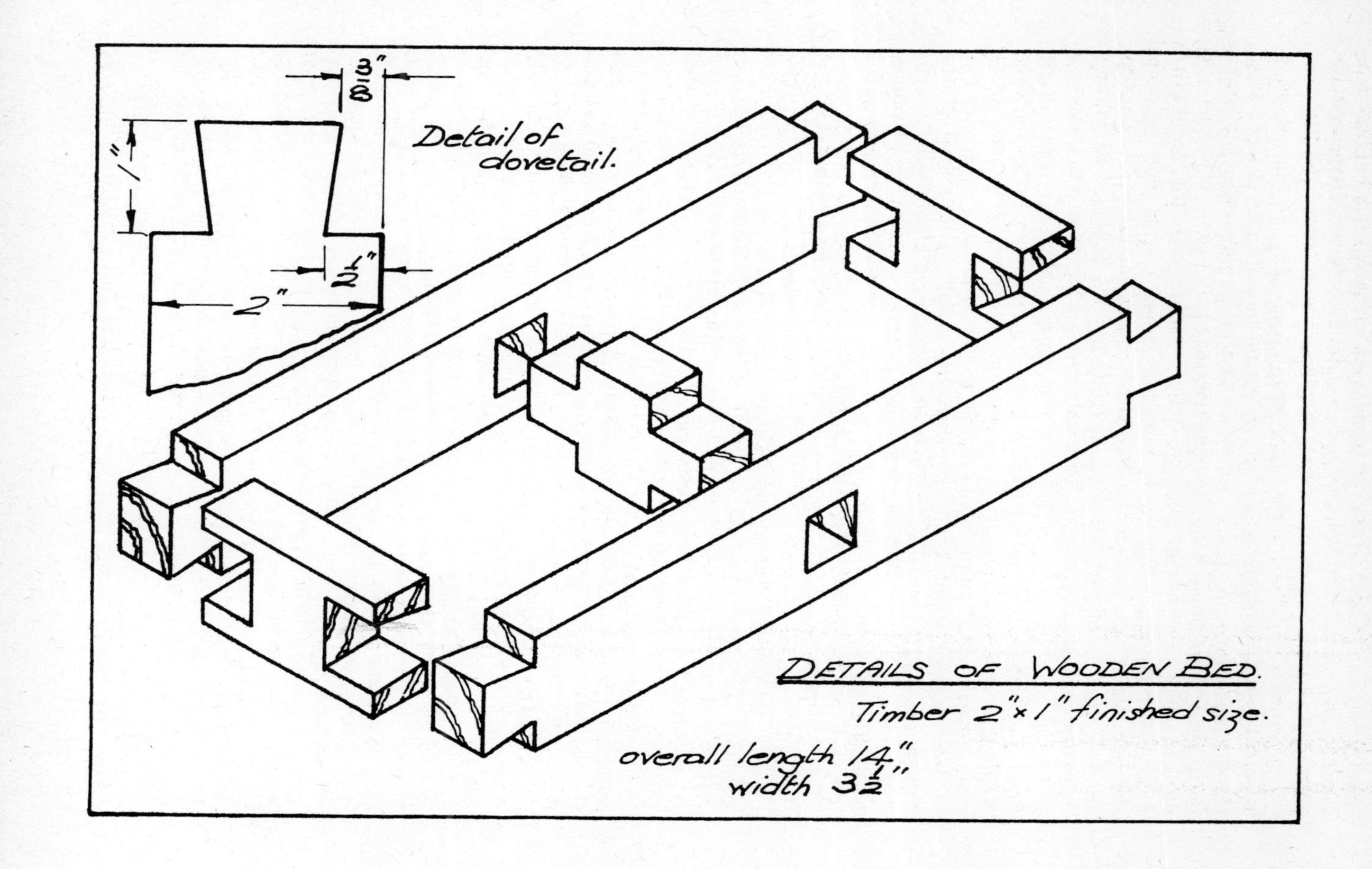
3/8"
1"
Detail of dovetail.
½"
2"
DETAILS OF WOODEN BED.
Timber 2"×1" finished size.
overall length 14"
width 3½"

plummer blocks and cylinder brackets, light brown. This latter, I have now decided, I don't like. Whether to do them dark green like the rest of the engine or a contrasting colour is something on which I just cannot make up my mind. Decisions, decisions, always decisions!

Clean off any oil, cutting or tapping fluid, etc., from the surfaces before painting or they may stain the finish. How many undercoats and topcoats and how much rubbing down you do is something you must decide. If you are a glass-case fanatic you will already know what is required. If, as I imagine, you are new to the game, then a good rub down over the surfaces to be painted with some fine emery cloth, two coats of undercoating, rubbed down lightly; then a top coat, which must be given ample time to dry thoroughly, then very lightly rubbed down, then the final top coat, All this should be done in a room which is reasonably dust-free. Surprisingly enough, the workshop—because there tends to be quite a bit of oil about to trap the dust and because when you are not there no-one is likely to be moving about in it—is probably the best place. An added protection is to make a wire frame, from spare clothes hangars, to stand over the engine and over which newspaper may be laid.

While the paintwork is drying, make the paper gaskets for the cylinder and steamchest. These should be drawn out accurately on stout brown paper—Stuart Turners sell a special gasket material for this purpose. Cut out and make the holes. I use a leather punch, but it can be done by assembling the gaskets in place and pushing, with a screwing action, the correct size of screw through the hole.

Now for final assembly. The engine soleplate is bolted to the wooden bed; machine off the bolts so that then fully tightened the domed end of the bolt just shows proud of the nut. Then screw the floor to the base. Assemble pedestal, bearings and crankshaft. Adjust so that crankshaft runs smoothly, remember to allow for the fact that when the bearing pedestal is bolted down to the floor, the wood will "give" a bit. Be prepared to have to dismantle the whole assembly to plane a little from the underside of the wooden bed, or to add shims, in the form of paper between the bed and the floor or brass/steel under the pedestal. One could go on for ages explaining what to do, but it is just wasting paper; the whole point is to keep adjusting and testing until it is absolutely right. Even if you spend a whole evening on it you will be grateful in the end, otherwise the cylinder will be developing power that is being mopped up by friction in the crankshaft. If it is a "stick-slip" situation, the engine will never run smoothly at slow speeds. It is easy to make an engine run fast, to get it to run slowly *and* smoothly is a lot more difficult. Bolt down the pedestal using $\frac{1}{4}$ inch BSF countersunk-head set screws and nuts.

Now replace the crankshaft complete with eccentric sheave, and flywheel. Have the outer bearing close up to the flywheel with any extra shaft length protruding ready to take a belt pulley, about which more later. The connecting rod can also be fitted. Make reference to the general arrangement drawings for the position of the various components.

If, from your initial assembly, you know that all the cylinder parts are satisfactory, final assembly of the cylinder, complete with gaskets and graphited packing on piston and in glands, can now take place—

except for the steam chest cover. The piston and slide valve will be noticeably stiffer to move, but from your previous assembly you will know that this is only due to the gland and piston packing. Needless to say, this must not be so tight that it requires any great effort to move them!

Bolt the cylinder assembly to the soleplate but do not tighten down, and attach the piston rod to the connecting rod with the wristpin. There should be 1/32 inch clearance between the piston and cylinder cover at each end of the stroke. Turn the engine over via the flywheel and evenly tighten down the cylinder mounting bolts as you do so.

Put the crossheads on the ends of the wristpin. Then assemble the lower guide bar supports and the lower guide bars in place and pass the long 7BA studs through them and screw into the soleplate. Check that with finger pressure on the top of the crossheads, the engine still turns over freely. Any adjustment will be by reducing the lengths of the $\frac{1}{4}$ inch diameter supports or alternatively shimming with metal or paper shims. Now put the $\frac{3}{8}$ inch long spacers on the studs and add the top bars. Again with slight pressure on the bars see that the engine rotates freely and, if necessary, adjust accordingly.

Assembly of the valve operating mechanism starts by pressing the lever (32) on to one end of the bridge rod (33), then assemble the rod to the bridge (40) passing it through the second lever (30) in the process. Set the two levers diametrically opposite each other and drill a 1·55 mm or No. 53 hole through this lever (30) and the rod. Cut a 5/16 inch length of 1/16 inch diameter steel wire, taper the end slightly with a smooth file in the lathe and tap the wire through the hole in lever and rod like a taper pin. Taper pins of this size may be purchased commercially, but as you will probably have to buy a quantity in order to get one it hardly seems worth while.

Mount the bridge assembly at the front end of the guide bars and nut down all round. The 7BA nut that comes under the rod is a bit awkward, but a little patience and a 7BA open ended spanner will soon put it right.

Mount the eccentric rod assembly on its sheave and connect the fork and lever at the front end with a pivot made as shown in the detailed sketch of these items. Now join the top of the lever (30) to the end of the valve rod by means of the links (29). As the top of the lever describes an arc while the end of the valve rod has to move in a straight line, these links, although simple, have the important duty of allowing a circular motion to change into a straight line motion.

After the links have been fitted, restrain the mad impulse to spin over the flywheel or you may find that the valve will bring everything to a sudden halt by thudding into the end wall of the steam chest! Instead slowly turn the flywheel over, in a clockwise direction, with the engine viewed from the side and the cylinder on your left, and adjust the position of the slide valve by undoing the link to valve-rod joint and rotating the valve rod thus screwing it back or forward through the valve-nut. When the valve exposes the steam ports by the same amount at each end during one rotation of the flywheel, the position of the valve, relative to the ports is correct.

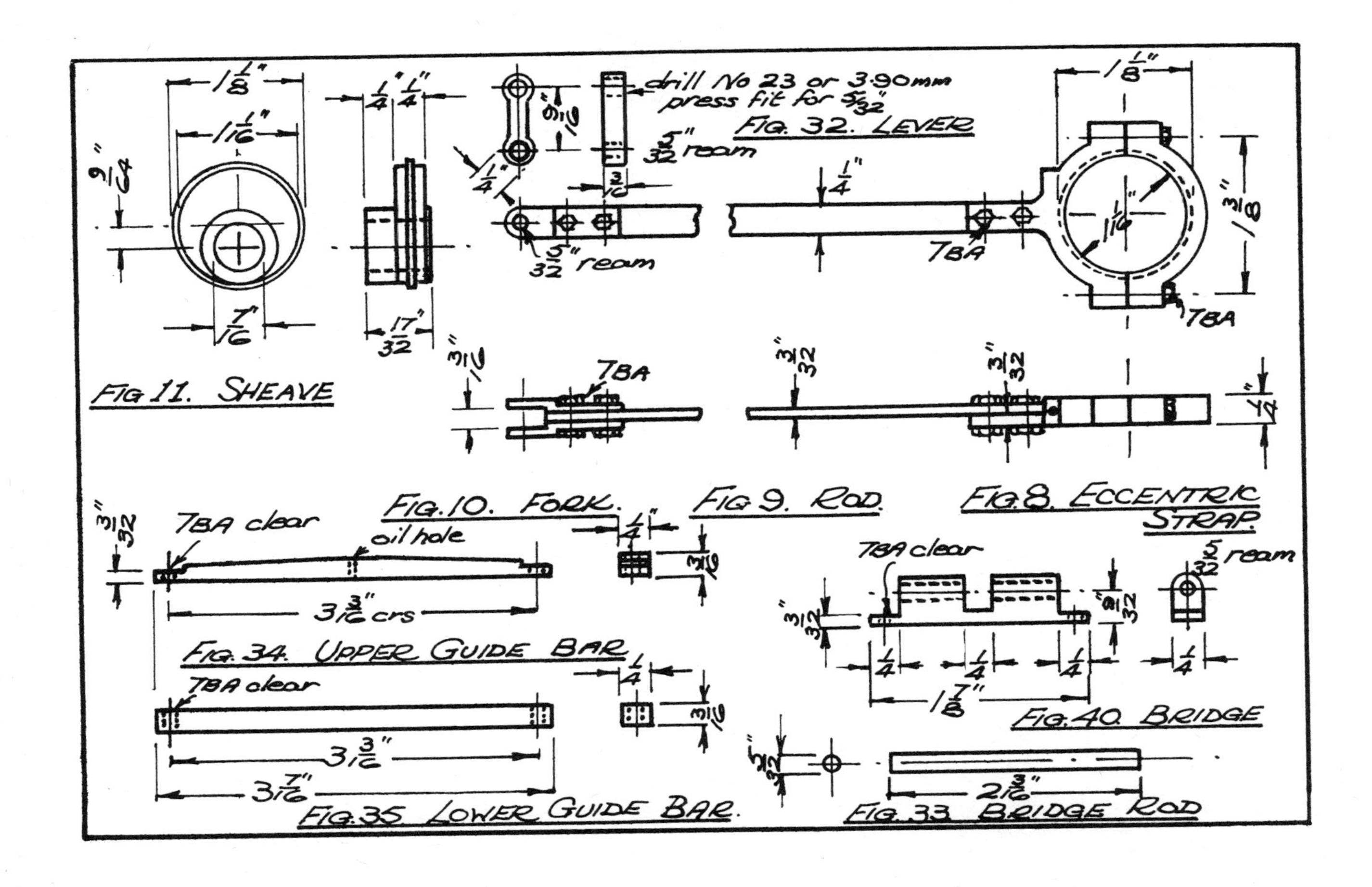

FIG 11. SHEAVE
1 1/8"
1 1/16"
9/64"
7/16"
1/4" 1/4"
17/32"
drill No 23 or 3·90mm press fit for 5/32"
FIG. 32. LEVER
9/16"
5/32" ream
1/4"
3/16
5/32" ream
1/4"
7BA
1 1/8"
1 3/8"
1 1/16"
7BA
3/16"
7BA
3/32"
3/32"
1/4"
FIG.10. FORK.
FIG 9. ROD.
FIG.8. ECCENTRIC STRAP.
3/32"
7BA clear
oil hole
3 3/16" crs
1/4"
3/16
FIG. 34. UPPER GUIDE BAR
7BA clear
1/4
3/16
3 3/16"
3 7/16"
FIG.35 LOWER GUIDE BAR.
7BA clear
5/32 ream
3/32"
3/32
1/4
1/4
1/4
1/4
1 7/8"
FIG.40. BRIDGE
5/32"
2 3/16"
FIG. 33 BRIDGE ROD

The next adjustment is to relate the position of the valve to the position of the piston, in other words, set the valve timing. Slacken the grub-screw which locks the eccentric sheave to the crankshaft. Turn the flywheel over in the direction of running, and stop when the piston reaches dead centre position, i.e. piston at the end of its stroke. Now rotate the slackened eccentric sheave, in the same direction and carefully watch the slide valve. Let it pass the port at the same end as the piston is positioned and, as it comes back, as soon as it starts to open the port, stop turning the eccentric and lock the grub screw. Continue turning the flywheel and check that the tiny port opening (lead) is the same at the other dead centre position. If it is not so, split the difference between each end. If the error is considerable, check all dimensions.

With the valve set, the valve cover can be bolted down not forgetting to fit the oiled paper gaskets. Now you can sit back and admire your handiwork.

The use to which you are going to put the engine will, to some extent, decide what final bits and pieces will be needed. If it is to repose on the mantlepiece, little else will be required apart from a list of instructions to the lady of the house setting out the necessary cleaning techniques. If, as mentioned earlier, you envisage a working model depicting an engine house including boiler, shafting, stone or brick walls, tools, etc., you will require a few more bits and pieces. A driving pulley will be needed on the end of the crankshaft. This could be turned from the beam engine pulley but must be for flat belt, i.e. slightly crowned, not for individual ropes as on the beam engine. Alternatively the pulley used on Mr. Taylor's Undertype Engine might be more in keeping. The drive on our full size original was via a 4 inch wide flat belt, hence a leather belt about $\frac{3}{4}$ inch wide would be about right on our model.

The steam valve No. 153/2 can be fitted above the steam chest to act as stop valve with another valve at the boiler to act as shut off. The correct type of boiler for such a layout would be a coal-fired vertical of approximately 6 inches diameter by 12 to 14 inches high, less chimney of course. The exhaust pipe might be turned downwards to pass through the floor or out through the wall, if the engine was mounted in the corner of the room.

One does not often see this type of working "scenic" model; an attractive example is Mr. Burgess's saw-mill (SIMEC Newsletter No. 8/1973) and it appeals to my simple mind much more than a gory battle scene!

You may have noticed the omission of lagging on the engine cylinder. The reason is simple, there was none nor evidence of there ever having been any, on the original. Perhaps one day I'll get round to fitting wood lagging with brass bands, which would look rather nice. Item 17 in George Watkin's book "The Stationary Steam Engine," shows the method of fitting and gives some good ideas for setting up this type of engine.

Well, that brings us to the end of the story about the building of "Victoria." I hope you have enjoyed reading it; even more I hope you enjoy building this model of the type of engine that drove the machinery in such numbers of small manufactories throughout Britain, that we earned the title "workshop of the world."

6. Developing the "Victoria" Engine

It has given me great pleasure to see how popular this engine has become. I suppose that to any steam enthusiast the horizontal stationary engine is a traditional type that we all understand and appreciate. The proportions of this particular model make it ideally suitable both as a working version and as an example of mechanical art to grace the sideboard.

At the 1980 Model Engineering Exhibition there were no less than five examples of this engine on show, showing embellishments and a quality of craftmanship that I can only envy. I believe at least three of them gained awards.

The first development of this design that has caught my attention is the appearance of a "twin" version. The general arrangement views of the Stuart Turner double "Victoria" illustrates how the design may be modified to become a model of a large and powerful engine for driving a factory or mill via multiple rope drive pulleys and lineshafts.

Fellow steam enthusiast and SIMEC President, Tom Walshaw, has built two magnificent examples to his own designs. One of them, called "Princess Royal", represents a 360 hp textile Mill engine, while the second named "Goliath" depicts a 1,200 hp cogging mill engine as would have been found in a steel works.

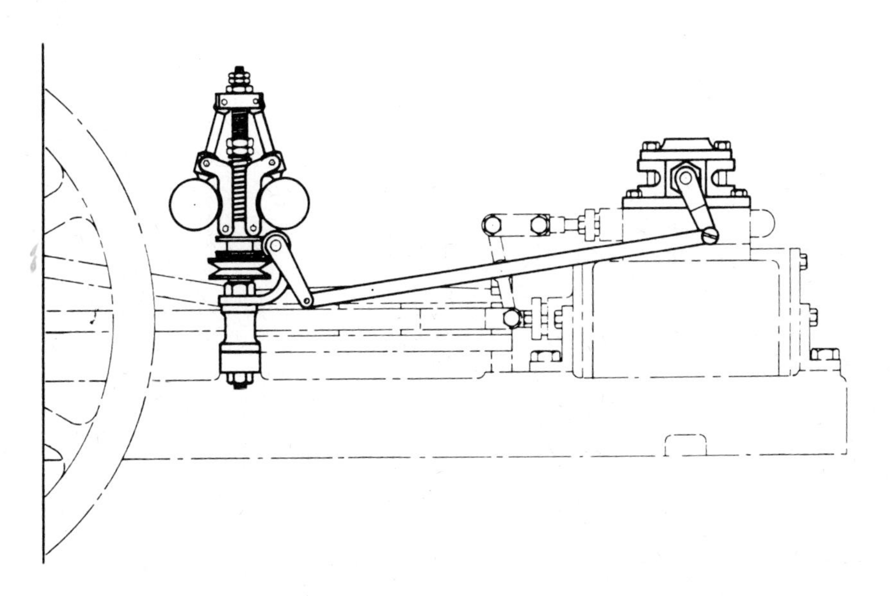

Diagram showing the details of the Stuart Turner Governor for "Victoria".

The standard single cylinder "Victoria" fitted with Governor.

(Photo: T. D. Walshaw)

A model, to a scale of one-twelfth full size, of a Clayton & Goodfellow "Pusher" Textile Mill Engine. It was designed to exhaust to an existing Beam Engine at Lowerhouse Mill, Burnley. Built by T. D. Walshaw and named "Princess Royal" it obtained a VHC Diploma at the 1980 Model Engineer Exhibition.

"Goliath". Two views of an 18 inch bore by 36 inch stroke two-cylinder Cogging Mill Engine. Built to a scale of 1 *to* 18 *by T. D. Walshaw. This engine was awarded a VHC Diploma at the Model Engineer Exhibition.*

(Photos: T. D. Walshaw)

These two models admirably demonstrate how the standard "Victoria" castings and layout can be redesigned and built up into examples of engines representing vastly different scales. That of the original "Victoria" being one-fifth full size while "Princess Royal" is one-twelfth and "Goliath" one-eighteenth full size.

Construction of the engine in its twin form should pose no great problems to the builder, beyond the extra work entailed. However, there are a couple of points that may be worth mentioning. First, due to the fact that the crankshaft is overhung at both ends and common to both engines, it is important that the centre heights of each engine be carefully checked. It is possible, of course, to "shim-up" one of the engine units so that they are in true alignment with each other, but this savours of "bodging", and I am sure no self-respecting steam engine constructor would be happy with the knowledge—even if not made public—that his engine relied on a couple of sheets of notepaper for its free-running qualities!

The second feature is that the crankshaft should have its crankwebs fitted so as to be at 90° to each other. This results in there being four power strokes during every revolution of the crankshaft which makes a very smooth running engine as well as one that is self-starting.

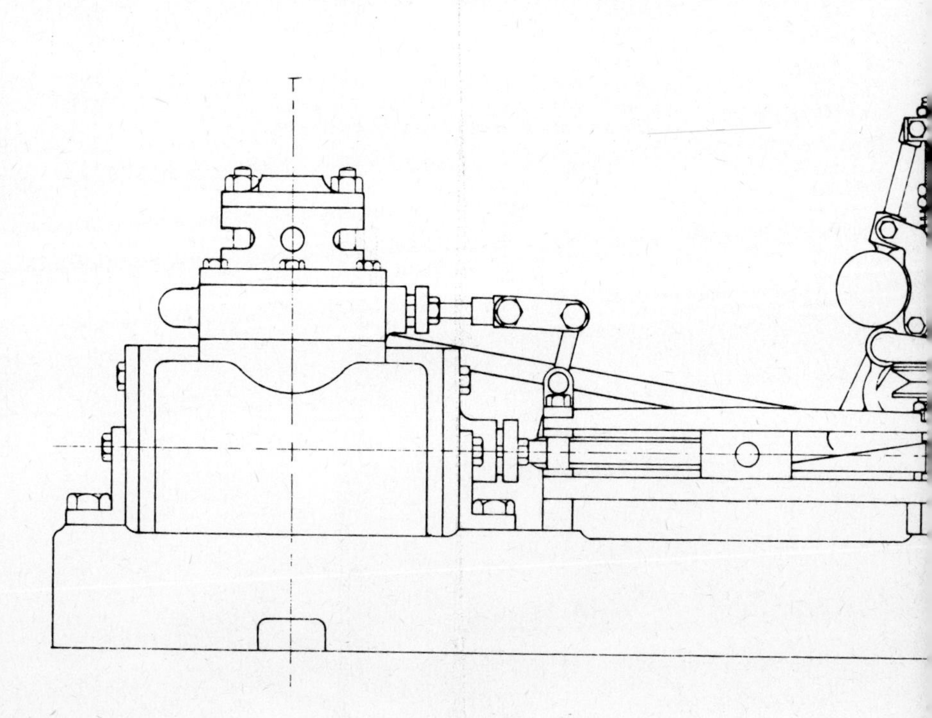

While dealing with the crankshaft, please note that the eccentric sheaves, flywheel and governor pulley (if fitted) must be correctly mounted on the shaft before the second crankweb is finally fitted. The same applies to the main-bearings if they are not split.

In addition the engines comprise a right- and left-hand pair and this must be kept in mind during construction and assembly.

Most stationary steam engines were fitted with governors, the purpose of which was to control the speed of the engine as the power requirements within the factory or mill altered. In general these governors could be divided into two types. The first and earliest controlled the engine speed by throttling the steam supply to the engine; the second method, was to vary the cut-off, by altering the travel of the valve. This, of course, pre-supposed that the engine was fitted with some form of variable valve gear. The form of governor that Stuart Turner Ltd. have developed for the "Victoria" engine is of the steam throttling type, its initial design being due to no less a person than James Watt himself.

General arrangement drawings of the twin "Victoria" now available from Stuart Turner Ltd. as a detailed set of castings, materials and drawings. Note that only one governor is required to control the pair of engines.

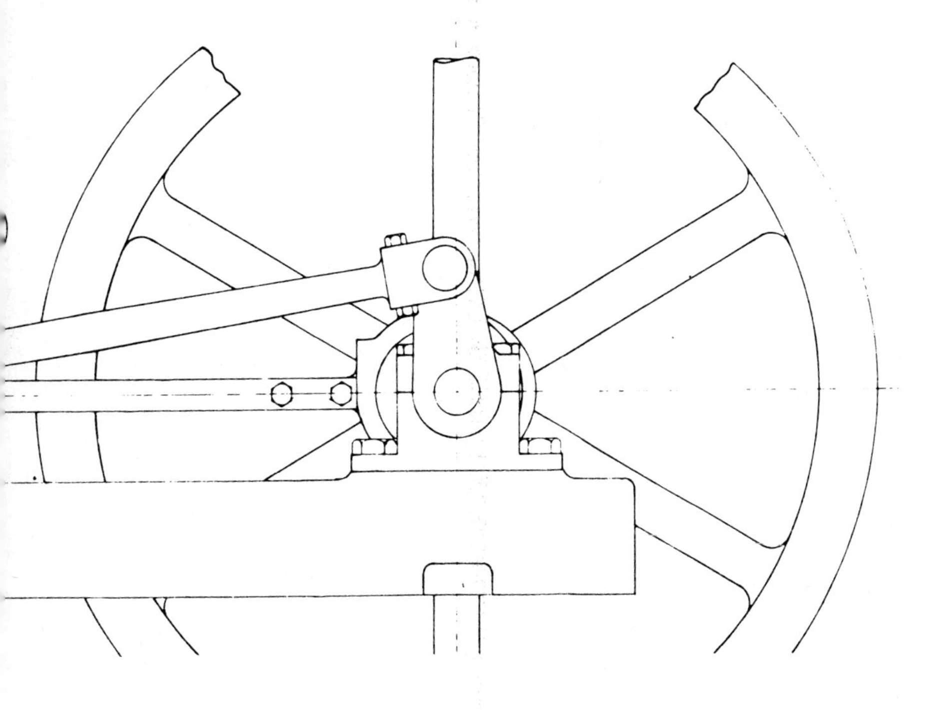

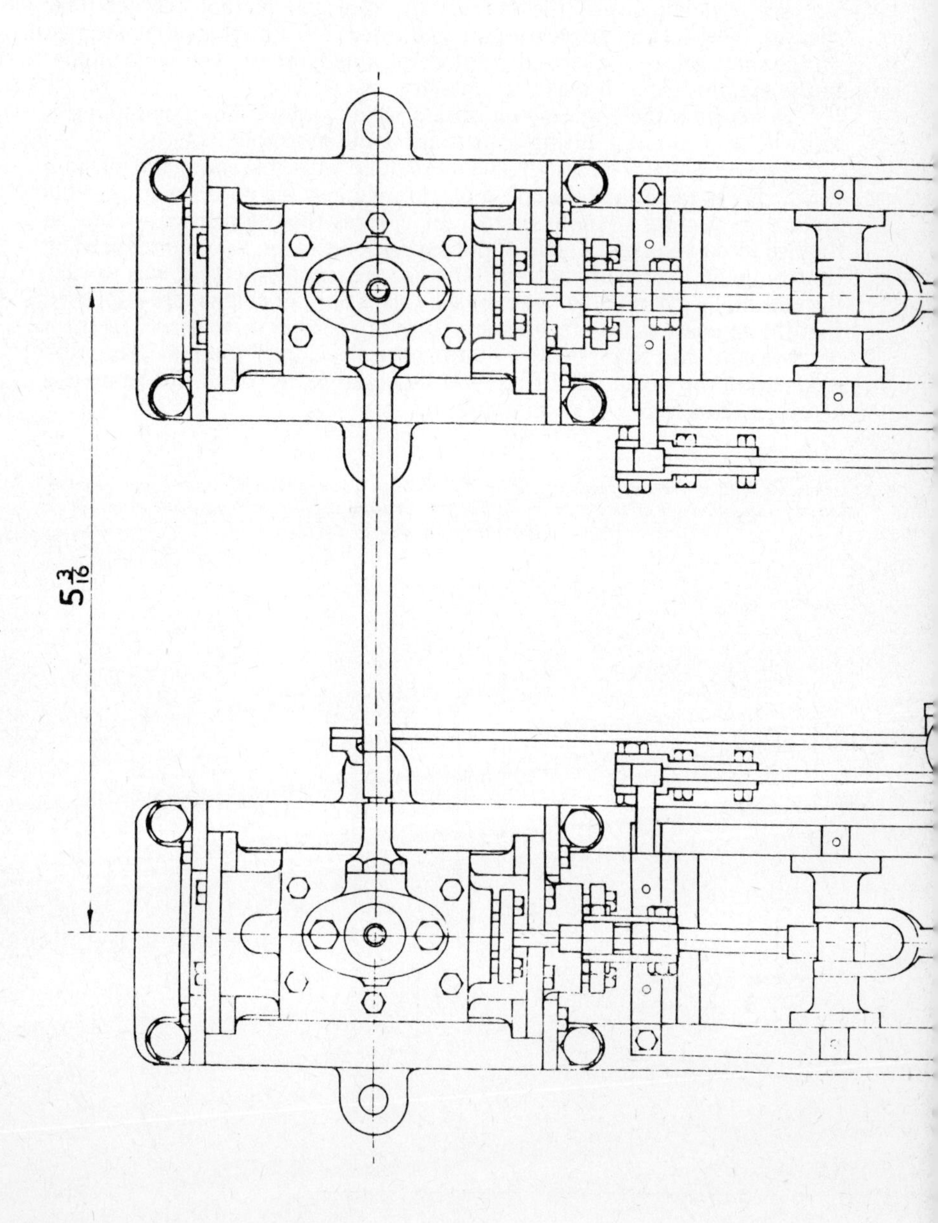
5 3/16

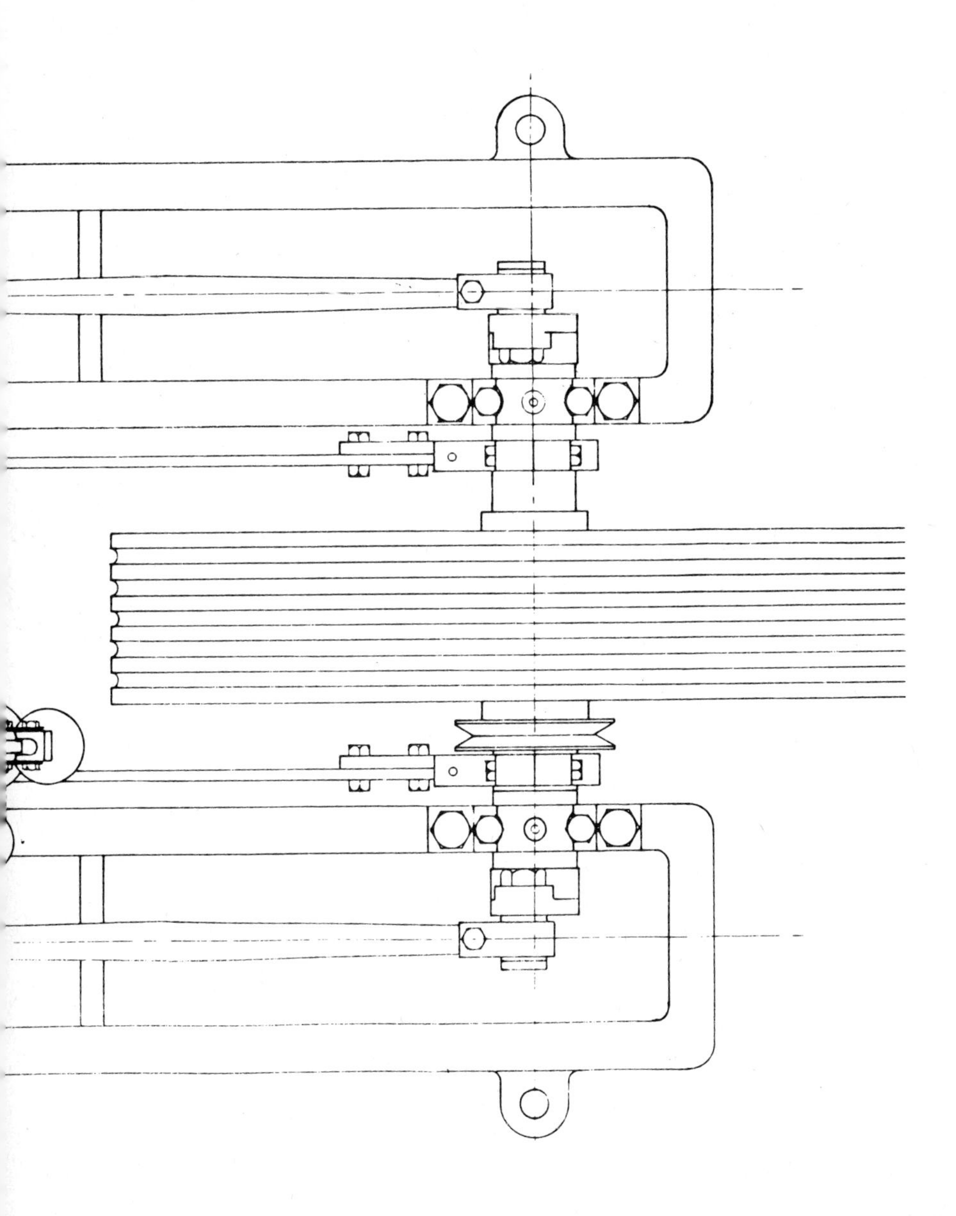

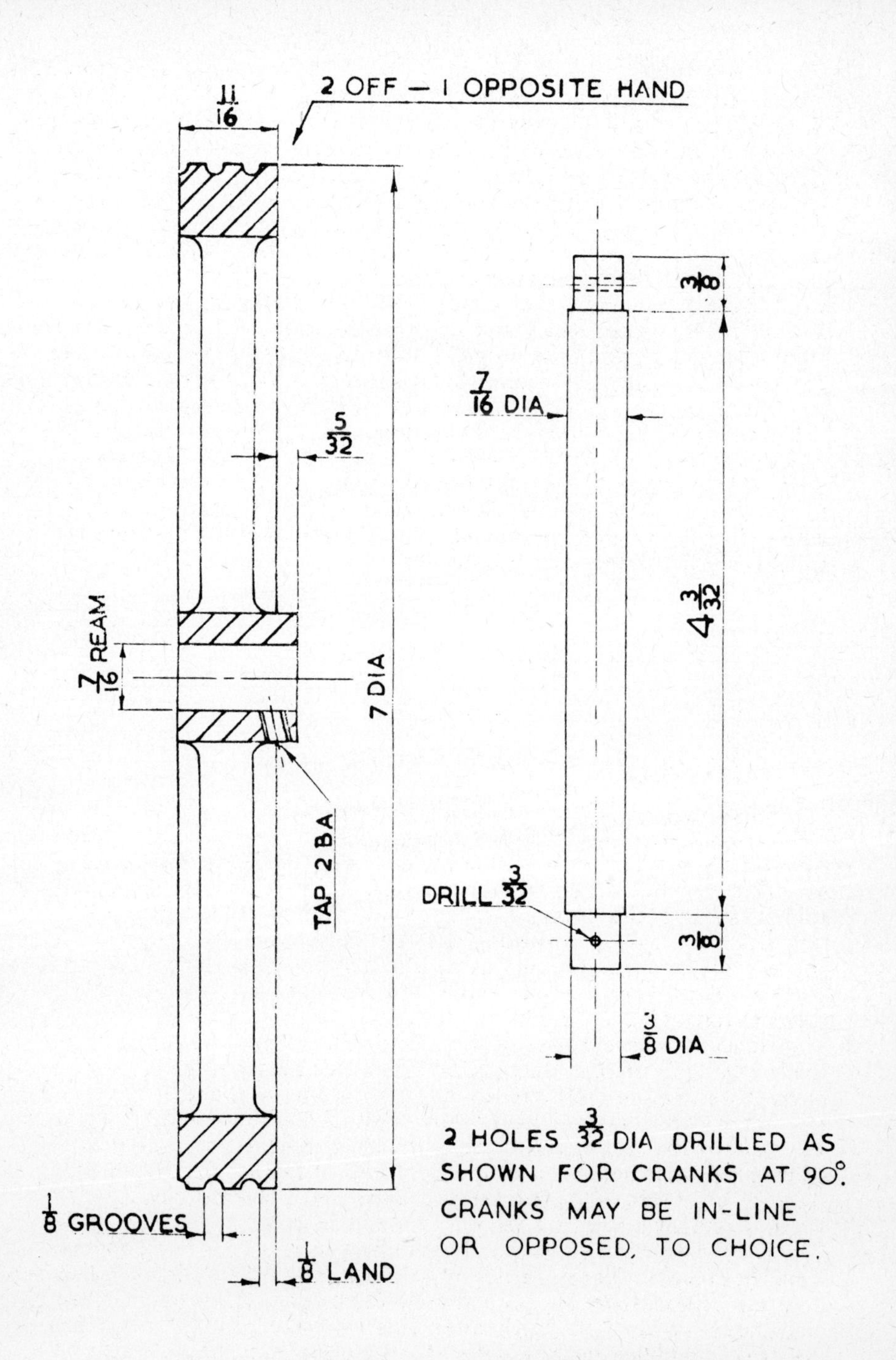

Details of crank-shaft and grooved flywheel for twin "Victoria".

Referring to the diagram that appears in this chapter, the operation of the governor is as follows. The governor pulley-spindle which carries the two flyballs is driven by a belt from the engine crankshaft. As the governor assembly rotates, the balls fly out under the action of centrifugal force, and, in doing so, raise a collar, to which is attached by a linkage a butterfly throttle mounted above the steam chest. The governor assembly is spring loaded so that as the speed of the engine falls, the flyballs drop, and so allow the throttle to resume its open position.

Although the individual components that comprise the governor assembly look rather small and complicated, the construction of this additional item should pose no great difficulty to the individual who has previously completed the engine. Stuart Turner Ltd., as is their custom, supply detailed drawings and a full set of the parts necessary to complete the governor. The only item that I would bring to the notice of prospective builders, is the spindle pulley, which is listed as Item No. 2. This requires a long hole of small diameter to be drilled through its axis, and is required—to run on Item 1, the spindle. This, I would suggest, being an example of where the hole through the spindle pulley should be drilled first and the spindle itself turned to form a suitable fit.

7. A Double-Size "Victoria" Mill Engine

Shortly after this book was first published, I had the good fortune to meet an engineer who was also a missionary with the African Inland Mission. He was very interested in the possibility of simple steam engines as power sources, particularly for pumping water, in remote villages. He told fascinating tales of his efforts at building primitive beam engines for well pumping and of converting a model petrol aeroplane engine to a single-acting steam engine and using it to drive a small fishing boat supplied by steam from a copper kettle! We even got on to considering the calorific value of camel dung!

He had obtained a copy of the "Victoria" book and was considering the feasibility of using an enlarged version of the design as the basis of an engine that could be built by the young Africans in his charge.

The photographs and details show the double-sized "Victoria" which was built by the missionary's brother-in-law, also an engineer, as a result of their deliberations. Most of the work was of fabricated construction, the base being welded up from steel channel and angle sections. The fly-wheel is actually a cast iron belt pulley found as scrap.

The construction of the cylinder is of particular interest being built up from the standard Stuart Turner 5A cylinder. This cylinder is of $2\frac{1}{4}$ inch bore and 2 inch stroke. The bore was reduced by press fitting into it a liner machined from cast iron (Meehanite). This was made 5 inches long so that the stroke could be fixed at 4 inches. Turned pieces were fitted at each end along which passage-ways were cut to act as steam-ways. The cylinder

The crankshaft end shows details of bearing mountings and big-end. The base-plate is built up by welding from channel and angle steel section.

The cylinder end of the large "Victoria" mill engine. All the details are built up from bright mild steel bar. A displacement lubricator is fitted and drain cocks are controlled by the linked handle on the lower left.

cover studs are extra long, passing through these new end pieces and into the original cylinder block. These annular extension pieces were made from mild steel so that the feet, made from steel angle, could be welded to them. In this way the scheme differs from the original "Victoria" design.

Drain valve extensions are fitted to take cocks which are below the cylinder in the base of the engine. The photograph of the cylinder end shows how they are linked to operate together from one lever. A large displacement lubricator is fitted to keep this powerful and hardworking engine well supplied with steam oil.

There is a great satisfaction in running and caring for a sizeable steam plant, it gives a pleasant atmosphere in a model engineering workshop and can earn its keep by driving a dynamo and/or lineshaft. While the basic "Victoria" design could be considerably simplified to provide a more robust utility steam engine, nevertheless the technique used by the builder of this double-size version to produce the traditional type of long stroke cylinder suitable for a horizontal engine, has much to recommend it.

A side view of the double-size "Victoria".

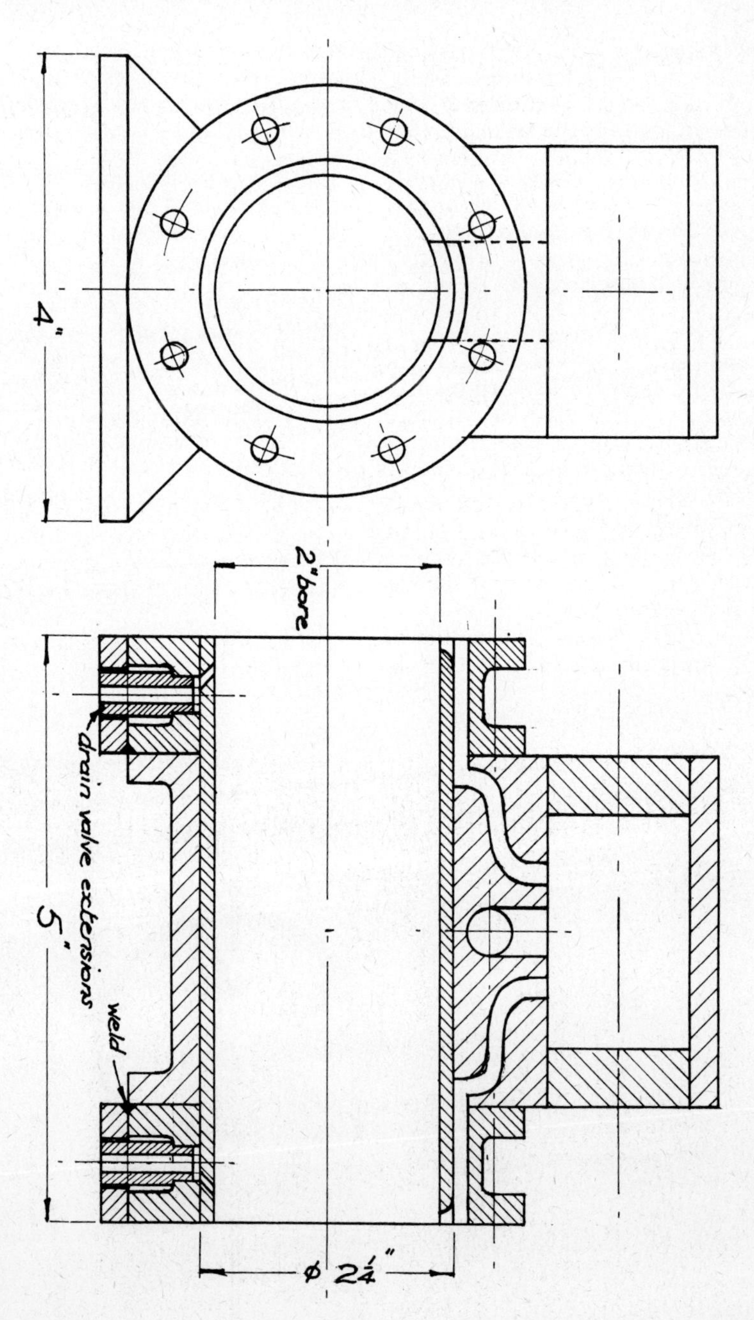

This drawing illustrates how a Stuart Turner No. 5A cylinder block may be sleeved and modified to form a double-size, 2 inch bore by 4 inch stroke, block for a "Victoria" built to twice size dimensions.

8. Camden Industrial Museum

The firm of J. B. Bowler & Sons Ltd. finally ceased to do business and closed its books in 1969. And it was with a heavy heart that, during the early 1970s, I watched the demolition gangs move in and raze this fascinating example of Victorian industry to the ground.

Little did I know at the time however, that an Industrial Trust had been formed in Bath, with the expressed purpose of saving as much as possible of the artifacts and interior fittings of the Bowler works, with the intention of using it as a basis to show the way in which industry had developed over the past 150 years or so in Bath.

The city has, for a long time, been noted for, amongst others, such illustrious names as Beau Nash, the Woods and Jane Austen, so this "rude mechanic" was delighted to learn in 1978 that a museum was to be opened to show, by means of the items that had been rescued in the nick of time from Bowler's Works, the multitude of small manufactories that had existed in the back streets and mews of this elegant city.

The museum gives honour to the foundryman who cast intricate brass items for local breweries and laundries; the expert bell-hanger who was there when needed, and the fettler, who started work at 4 every morning cleaning the castings poured the day before, ready for the brass finishers who arrived at 7 a.m.

My first visit to the museum proved to be a most pleasant surprise. My expectation was to find lots of shelves and glass cases full of the various

"Bowler's Shop". This will be recognised by many generations of Bathonians as the place where something could always be found to do that awkward job.

Line- and counter-shafts driving Victorian machinery. First from the original of "Victoria", and later from gas and oil engines, none of which were ever removed from the premises until they were finally demolished in 1970.

Part of the works just before being demolished.

The photographs in this chapter are by courtesy of Bath Industrial History Trust.

items rescued before the demolition of the works. Instead, I walked through the door to find myself in the actual shop! There was the counter I had leant on ten years before, spending many hours talking steam to the last of the Bowler proprietors. The counter flap was invitingly open and, on walking through into the back of the "shop", found myself in Mr. Bowler's own office, complete with fireplace, cupboards, hanging slates (on which to "put it"). From the Office a few paces took me into the Pattern Shop complete with wood-turning lathe. Then from here into the Brass Foundry where most of Mr. J. B. Bowler's seven sons worked. From here I moved into the Brass Finishing Shop, full of treadle lathes that would today earn a fortune in an antique sale.

Also on view, and working, is the General Engineering Shop, full of lathes, shaping and milling machines; every one an antique in its own right. I only regret that I no longer own the original of "Victoria" so that I could give it to the museum to drive this exhibit instead of the concealed electric motor at present used.

J. B. Bowler & Sons Ltd. were also makers of mineral waters, a fact to which I testified in the first chapter of this book. The museum therefore, also contains a mineral water factory, complete with Essence and Syrup Room, Bottle Store, Filling Area and Office.

I am happy that the original "Victoria" model is now on loan to the museum and I hope that readers will, when in the West Country, make a point of visiting this fascinating re-creation of a long-gone scene of craft and industry.

Some relevant details are that the museum is situated at Camden Works, Julian Road, Bath, and that it is open daily, except Friday, from 2 to 5 p.m.

Other books in the Stuart Turner series devoted to model steam engines, all by the same author:

BUILDING THE "JAMES COOMBES TABLE ENGINE"—
constructional details.

BUILDING THE "OCLE", OPEN COLUMN LAUNCH ENGINE.
An excellent beginner's book showing how an open chamber launch engine can be built easily from parts and castings.

BUILDING A "REAL" VERTICLE STEAM ENGINE.
How to build an elegant 19th century engine from Stuart parts. Text, scale diagrams and photographs.

BUILDING THE "STUART" BEAM ENGINE.
A practical guide for the novice or expert of how to build the popular and spectacular Beam Engine from Stuart parts. Fully illustrated.

BUILDING A VERTICAL STEAM ENGINE FROM CASTINGS.
Complete instructions for building a Stuart IOV engine with details on machining bedplate castings, trunk column and cylinder, crankshaft piston, conrod and valve turnings.